Lifetime Estimation of Welded Joints

Tadeusz Łagoda

Lifetime Estimation
of Welded Joints

Tadeusz Łagoda
Politechnika Opolska
Katedra Mechaniki I Podstaw
Konstrukcji Maszyn
45-271 OPOLE, ul. St. Mikolajszyka 5
Poland
t.lagoda@po.opole.pl

ISBN: 978-3-540-77061-9   e-ISBN: 978-3-540-77062-6

Library of Congress Control Number: 2007940901

© 2008 Springer-Verlag Berlin Heidelberg

This work is subject to copyright. All rights are reserved, whether the whole or part of the material is concerned, specifically the rights of translation, reprinting, reuse of illustrations, recitation, broadcasting, reproduction on microfilm or in any other way, and storage in data banks. Duplication of this publication or parts thereof is permitted only under the provisions of the German Copyright Law of September 9, 1965, in its current version, and permission for use must always be obtained from Springer. Violations are liable to prosecution under the German Copyright Law.

The use of general descriptive names, registered names, trademarks, etc. in this publication does not imply, even in the absence of a specific statement, that such names are exempt from the relevant protective laws and regulations and therefore free for general use.

Cover design: WMXDesign GmbH

Printed on acid-free paper

9 8 7 6 5 4 3 2 1

springer.com

Preface

In the paper the author attempts to assess the fatigue life of chosen welded joints. It focuses especially on chosen problems that accompany determination of the fatigue life of welded joints, taking into consideration the strain energy density parameter. Chapter 2 describes the welded joint as a stress concentrator. The state of stress and strain in the notch are described and theoretical and fatigue coefficients are indicated. The fatigue coefficient of the notch effect is estimated on the basis of fictitious radius in the notch root. Chapter 3 presents a model of fatigue life assessment under uniaxial stress state with statistical handling of data presented. The new energy model of fatigue life assessment, which rests upon the analysis of stress and strain in the critical plane, is described in detail in chapter 4. The principle of such a description is presented in the uniaxial as well as in biaxial state of loading. Chapter 5 contains the analysis of tests of four materials subjected to different loadings: cyclic, variable-amplitude with Gaussian distribution, and variable amplitude with Gaussian distribution and overloading for symmetric and pulsating loading. The analysis is based on the determined fatigue characteristics for all the considered materials. Chapter 6 shows the application of the model in the fatigue life assessment in the complex state of loading (bending with torsion of flange-tube and tube-tube joints) based on fatigue research of steel and aluminum welded joints carried out in well-known German centres. Proportional and out-of-proportional cyclic research are carried out. Additionally, the influence of various bending and torsion frequencies and proportional and out-of-proportional variable amplitude loadings are analysed.

Dealing with such a complicated problem as fatigue life of welded joints is requires a wide cooperation with other researchers and research centres. That is why I would like to express his gratitude to at least some of the people who contributed to the issue of this publication. This book is a result of my research work, considerations and discussions while my six-months stay at LBF Darmstadt, Germany, financed by NATO. Thus, I would like to thank all the workers of LBF, especially Prof. C.M. Sonsino [218] and Dr. M. Küppers for many discussions, access for their test results, laboratories and library while my visit, and for information sent in our correspondence. I was able to complete my book while my later two-week stay in Darmstadt, financed by DAAD. In this book, I also used some data obtained from Technical University of Clausthal, Germany, namely from Prof. H. Zenner. I want to thank Prof. Zenner, Prof. A.Esderts and Mr. A.Ahmadi for their help while my one-week visit at TU of Clausthal, financed by CESTI.

I also used my experience obtained during my work with the postgraduate students at Opole University of Technology: Dr. Damian Kardas, Dr. Krzysztof Kluger, Ms Małgorzata Kohut, Dr. Paweł Ogonowski, Dr. Jacek Słowik and Ms Karolina Walat. I must also thank my co-workers from Opole University of Technology: Dr. Adam Niesłony, Dr. Aleksander Karolczuk, Dr. Roland Pawliczek, and especially Prof. Ewald Macha, also some other people not mentioned here. I would like to thank Ms Ewa Helleńska for translation of this book and some previous papers into English. Finally, I want to thank Prof. M. Skorupa and Prof. K. Rosochowicz for their suggestions and advice.

I wish to dedicate this book to my wife Bożena for supporting me in my research work as well as for her constant understanding and care.

Tadeusz Łagoda
t.lagoda@po.opole.pl
Department of Mechanics and Machine Design
Faculty of Mechanical Engineering
Opole University of Technology
ul. Mikolajczyka 5
45-271 Opole, Poland

Contents

Notation IX

1 Introduction 1

2 Welded Joints as the Stress Concentrator 5
   2.1 The Complex Stress State in the Notch 5
   2.2 Theoretical Notch Coefficient 8
   2.3 The Fatigue Notch Coefficient 10
   2.4 The Fictitious Radius of the Welding Notch 12
   2.5 The Notch Coefficient with the Use of the Fictitious
       Notch Radius 15

3 The Stress Model for the Assessment of Fatigue Life Under
   Uniaxial Loading 17
   3.1 Algorithm for the Assessment of Fatigue Life
       Under Uniaxial Loading State 17
   3.2 Statistic Evaluation 27

4 The Energy Model of Fatigue Life Assessment 33
   4.1 The Energy Parameter Under Uniaxial Loading 33
   4.2 The Energy Parameter Under Multiaxial Loading 39
       4.2.1 The Generalized Criterion of the Parameter of Normal
            and Shear Strain Energy Density Parameter in the
            Critical Plane 40
       4.2.2 The Criterion of Maximum Parameter of Shear and
            Normal Strain Energy Density on the Critical Plane
            Determined by the Normal Strain Energy Density
            Parameter 42
       4.2.3 The Criterion of Maximum Parameter of Shear and
            Normal Strain Energy Density in the Critical Plane
            Determined by the Shear Strain Energy Density
            Parameter 43
   4.3 Algorithm for Fatigue Life Assessment 47

5 An Example of Fatigue Life Evaluation Under Simple Loading 59
5.1 Fatigue Tests... 59
5.1.1 Tests Under Constant-amplitude Loading............................ 62
5.1.2 Tests Under Variable-amplitude Loading 66
5.2 Verification of the Results Obtained Under
Variable-amplitude Loading.. 67

6 An Example of Fatigue Life Evaluation Under Complex
Loading States 71
6.1 Fatigue Tests... 71
6.2 Verification of the Criteria Under Constant-amplitude
Loading... 80
6.2.1 The Parameter of Shear and Normal Strain Energy
Density on the Critical Plane Determined by the
Parameter of Normal Strain Energy Density........................ 80
6.2.2 The Parameter of Shear and Normal Strain Energy
Density in the Critical Plane Determined by the Shear
Strain Energy Density Parameter 86
6.2.3 The Influence of Different Frequencies of Bending
and Torsion on Fatigue Life.. 91
6.3 Verification Under Variable-amplitude Loading 92

7 Conclusions... 95

References... 97

Summary.. 117

Notation

A_5	contraction
C	coefficient containing circumferential stresses in the root of the notch
E	longitudinal modulus of elasticity (Young's modulus)
f	frequency
G	shear modulus
$k(N_f)$	ratio of allowable stresses for bending and torsion for a given number of cycles N_f
K_f	fatigue notch coefficient
K_t	theoretical notch coefficient
K_{ta}	theoretical notch coefficient for axial loading
K_{tb}	theoretical notch coefficient for bending
K_{tt}	theoretical notch coefficient for torsion
N_f	number of stress cycles up to fracture
m, m'	slope of fatigue S-N characteristic curve
M	moment
N	number of cycles
P	force
r	correlation coefficient
R	stress ratio
$R_{0.2}$, R_e	yield point
R_m	tensile strength
s	coefficient of multiaxiality, standard deviation
$S(T_o)$	fatigue damage degree in observation time T_o
t	time, thickness of sheet
T_N	scatter band for life-time
$\overline{T}_N$	mean scatter band for life-time
T_o	observation time
W	strain energy density parameter
α	angle of the plane position
γ	shear strain
ε	normal strain

X   Notation

φ	phase displacement angle
ν	Poisson ratio
ρ	radius in the welding notch root
σ	normal stress
τ	shear stress
0xyz	xyz co-ordinate system with the origin in the point 0

Indices

a	amplitude
af	fatigue limit
b	bending
cal	calculation
e	elastic
eq	equivalent
exp	experimental
f	fictitious
l	local
m	mean value
max	maximum value
min	minimum value
n	nominal
p	plastic
t	torsion
x, y, z	directions of axes of the co-ordinate system
w	weighed value

Functions

$$\text{sgn}(x) = \begin{cases} 1 & for \quad x > 0 \\ 0 & for \quad x = 0 \\ -1 & for \quad x < 0 \end{cases}$$

$$\text{sgn}(x, y) = \frac{\text{sgn}(x) + \text{sgn}(y)}{2}$$

1 Introduction

The problem of determination of fatigue life of welded joints has been investigated for many years. As a result, there is a possibility to find solution of that issue in many publications. Typical handbooks concerning fundamentals of machine building are for example [33, 194], other books and monographs [45, 53, 199, 203, 204, 215, 216, 234 and others] or the latest work [44], to mention some. The problem has been also discussed in many journal publications and presented in conferences. Only few of publications have been cited in this paper. During the last 15 years, the in-depth analysis of the problem of fatigue life calculations has been presented in many books and other publications [40]. The author of this monograph refers to the most important of them [4, 5, 40, 73, 100, 166, 176, 233].

Correct design of welded joints seems to be very important, for example in transport facilities, including hoisting equipment [84, 167] where special safety regulations must be fulfilled, or in the structures with high pressure of a medium [186].

According to [176, 232], in order to define the fatigue life in welded joints, there are two basic approaches possible to determine calculation stresses: first – on the basis of the nominal stresses, and second – on the basis of the strictly local stresses determined in the potential point of crack initiation ("hot spot").

Analysis based on the nominal stresses is applicable in the situation where the considered element has been classified and when the stresses can be easily determined. In [233], Susmel and Tovo presented satisfactory results of many calculations of welded joints based on nominal stresses under constant-amplitude loading.

The hot spot method is recommended for the cases where the strains can be measured near the joint [42, 176], or if the strains can be calculated with the finite element method. In [31] Dang Van et al., on the basis of the analysis of more than 200 fatigue tests of different steels (low- and high-strength) and different geometries of welded joints, founded that fatigue life of welded joints, calculated on the basis of stresses determined with the hot spot method [177, 178], is not strongly influenced by types of the materials joined. It is observable particularly for a number of cycles greater than $5 \cdot 10^5$. In the case of a lower number of cycles, for higher-strength materials, the permissible stresses are higher than for normal steels. It is important to draw attention to the fact that the notch coefficients for high-strength steels are greater than those for low-strength steels [224]. Thus, it is interesting to analyse relation between strength of normal steels and higher-strength steels in the local notation, i.e. including the theoretical notch coefficient. The safe fatigue life of the butt joints is higher than that of the fillet joints. Maddox [166] claims that good results can be obtained for nominal stresses but, in his opinion, the hot spot method should be developed in future. Principles of local stress determination according to the hot spot method are presented in Fig. 1.1. The local stresses can be defined from strains determined by extrapolation using two or three tensometers, or calculated with the finite element method.

If the local approach for welded joints subjected to multiaxial loading is applied, it is necessary to know the stress concentration for bending and torsion (K_{tb}, K_{tt}) at the fusion edge [165, 217, 220, 221]. Because of the fact that it is usually not possible to measure the actual radius of the fusion edge, in order to solve this problem, a suitable method is necessary. For welded joints subjected to uniaxial loading, the problem has been successfully solved owing to the application of so-called fictitious radius or, in other words, conventional radius [202, 208], based on the Neuber theory [173, 174]. Local methods for determination of fatigue life of welded joints under multiaxial fatigue were reviewed by Labesse-Jied [89]. The calculated fatigue lives of welded joints made of C45 steel and subjected to proportional and non-proportional random tension-compression with torsion loading were located in the scatter band of coefficient 4. The analysis was done on the basis of local stresses with plastic strains. The method using the conventional radius in the notch root can be applied for determination of the theoretical notch coefficient in

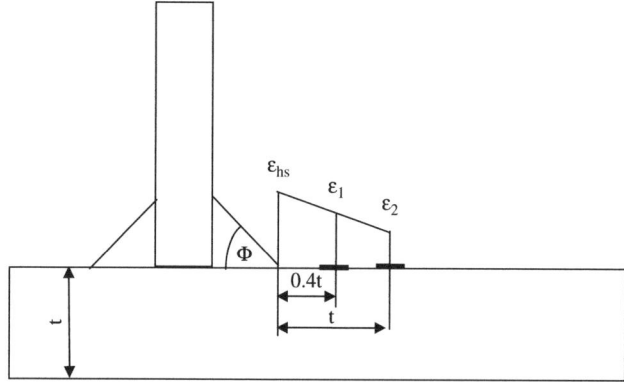

Fig. 1.1. Determination of strains with the hot-spot method

the case when the notch radius in the welded joint is small and tends to zero. In this paper, this method is considered for complex loading.

Similar to the Neuber's method based on the fictitious radius in the notch root, is the method proposed by Lawrence et al. (for example in [98]). In this model, determination of the maximum fatigue coefficient of stress concentration K_{fmax} is suggested. That value is determined for the critical radius in the notch root, equal to the critical value a* dependent on the material. It took the values from about 0.1 mm for welds made of high-alloy steels to 0.25 for low-alloy steels.

Another method of determination of geometric stresses, next applied for fatigue life calculations, was presented by Xiao and Yamada [246]. It was proposed by them to perform calculations with the use of stresses occurring 1 mm from the point of contact on the surface of the joined materials.

In [225] and [38, 47], Sonsino et al. claim that in practice damages should be accumulated on the assumption that the sum of damages according to the Palmgren-Miner hypothesis is $D = 0.5$. In [5], on the basis of the results of the tests under non-sinusoidal variable loading of the welded joints in bridges, at the drilling platforms or steel chimneys, it was observed that the sum of fatigue damages was $D < 1$. In [50], rough steel welded joints were subjected to variable-amplitude loading and it was found that damage accumulation in the considered joints varied about $D = 1$, and for the machined welded joints $D = 0.33$. In [96], Lahti found that the damage sum for variable-amplitude loading was less than 1. In [168], Mayer et al. stated that the experimental life was usually 3.5 times less than the life calculated according to the Palmgren-Miner fatigue damage accumulation hypothesis. Thus, it can be assumed that the damage sum is included in (0.33–1) according to various test results.

Standard recommendations referred to calculations of welded joints can be found in Eurocode 3 [38, 207], or the standards of the International Institute of Welding (IIW) [48] (for steels), and Eurocode 9 [39] (aluminium alloys).

In [12], typical fatigue diagrams for welded joints under axial loading (or bending) and torsion with constant inclination coefficients are presented (see Fig. 1.2).

In present paper two models of fatigue life estimation, based on stresses and the strain energy density parameter are presented.

For the uniaxial loading state (bending or axial loading), the model using local stresses was discussed. This model includes a value of the theoretical notch coefficient. As stated before, on the basis of [177, 178], it can be said that fatigue life of steel welded joints does not depend on a kind of material. In Chap. 3, there is a model of fatigue life assessment under uniaxial stress state with statistical handling of data presented. Chapter 5 contains the analysis of tests of four materials subjected to different

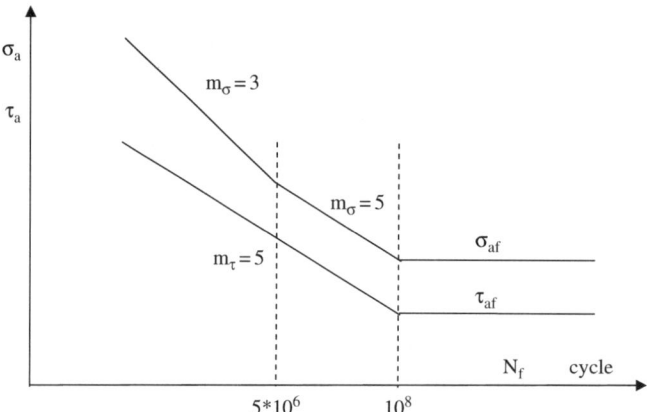

Fig. 1.2. Standard slopes of stress fatigue graphs for welded joints (σ_a – normal stress, τ_a – shear stress)

loading: cyclic, variable-amplitude with Gaussian distribution, and variable amplitude with Gaussian distribution and overloading for symmetric and pulsating loading. The analysis was based on the determined fatigue characteristics for all the considered materials.

Another approach is applied for the complex stress state. When the stress and strain tensors are determined for the welded joint, it is necessary to reduce the multiaxial loading state to the equivalent uniaxial state. For this purpose the fatigue effort criteria based on stress, strain, or the strain energy density parameter [115, 123, 136, 164] referred to the critical plane can be used. In Chap. 4, the energy model using the strain energy density parameter for complex loading was presented. This model includes both stresses and strains occurring in the material. In [156, 190, 193] it has been proved that in the case of great number of cycles the stress and energy models are the most appropriate for fatigue description, and for low numbers of cycles the strain and energy models are good. Thus, the energy model seems to be universal and it was verified many times in many papers concerning uniaxial loading [13, 14, 55, 59, 64, 65, 66, 83, 102, 103, 105, 106, 122, 136, 154, 179, 180] and complex loading [57, 83, 116, 118, 120, 124, 125, 130, 135, 140, 141, 142, 149, 150, 151, 152, 155, 187, 193]. In this paper, known results obtained for tube-tube and flange-tube joints under pure bending and torsion and their combination, in- or out-of-phase, and also for chosen steel welded joints under variable-amplitude loading [165, 217, 220, 221, 240, 241, 242, 243] and aluminum joints [86, 87, 88, 226] were evaluated. For analysis, some selected criteria based on the energy parameter for multiaxial fatigue were applied [32, 115, 123, 136, 164, 165, 214, 217, 220, 221].

2 Welded Joints as the Stress Concentrator

2.1 The Complex Stress State in the Notch

Complex stress concentration characterizes welded joints in which both geometrical and structural notches can be distinguished. In the case of geometrical notches under simple loading states, e.g. bending or axial loading, on the surface of the element in the notch root the plane stress state occurs. In round elements, apart from the nominal stress σ_x, the additional circumferential stress is also observed along the element. It can expressed by a formula

$$\sigma_y = C\sigma_x, \tag{2.1}$$

where $0 \leq C \leq \nu$. For simplification it can be written as

$$C = \begin{cases} 0 & \text{for } K_t = 1 \\ 0 \div \nu & \text{for } K_t \in (1,2), \\ \nu & \text{for } K_t > 2 \end{cases} \tag{2.2}$$

where K_t is the stress concentration ratio.

Analysing the results of calculations performed in LBF Darmstadt by Sonsino et al., the following equation can be formulated [193, 229]

$$C = \frac{1.84\nu}{K_t}(K_t - 1)^{1-\nu}. \tag{2.3}$$

The coefficient C, defining the values of circumferential stresses depending on the stress concentration ratio is shown in Fig. 2.1. From the analysis of the figure it appears that the value of the coefficient C tends to the Poisson ratio, ν, for K_t close to 2 (according to (2.2)).

It should be also noted that in the case of a sharp notch, the plane stress state is accompanied by the plane strain state (ε_x, $\varepsilon_y = 0$, ε_z). It results from the adoption of the elastic body model, the generalized Hooke's law, and $\sigma_y = \nu\sigma_x$ according to (2.1) and (2.2). For sharp notches, stress distributions for tension, bending and torsion are shown in Figs. 2.2, 2.3 and 2.4.

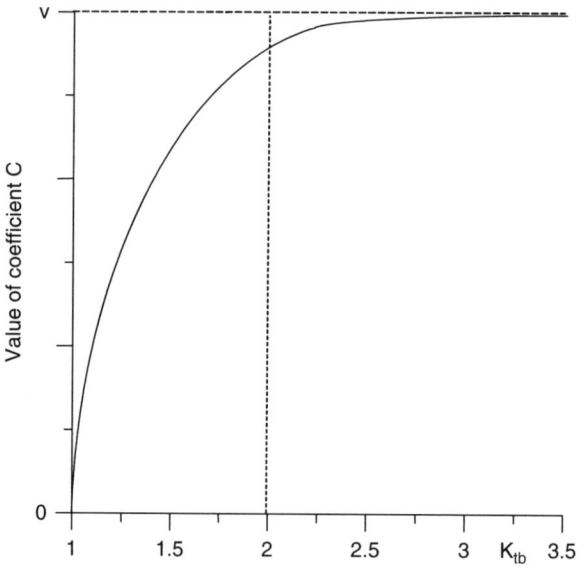

Fig. 2.1. Coefficient C versus theoretical stress concentration coefficient

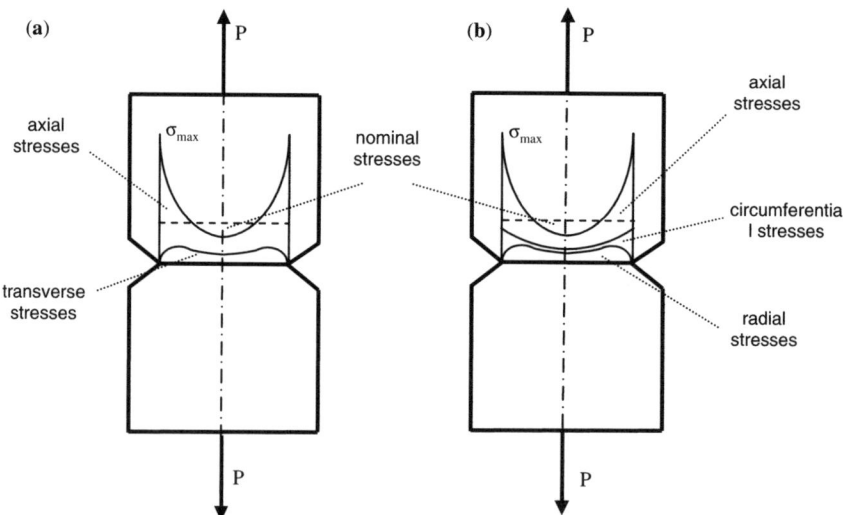

Fig. 2.2. Stress distributions in elements with sharp notches under tension: (**a**) flat element, (**b**) cylindrical element

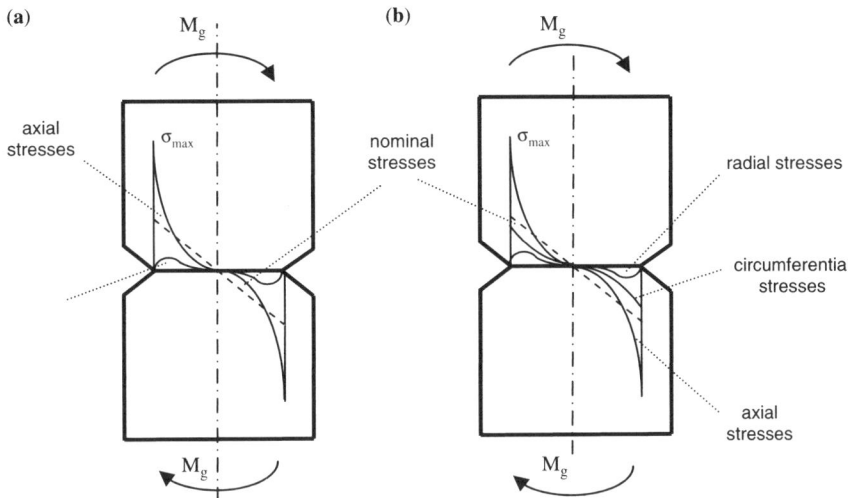

Fig. 2.3. Stress distributions in elements with sharp notches under bending: **(a)** flat element, **(b)** cylindrical element

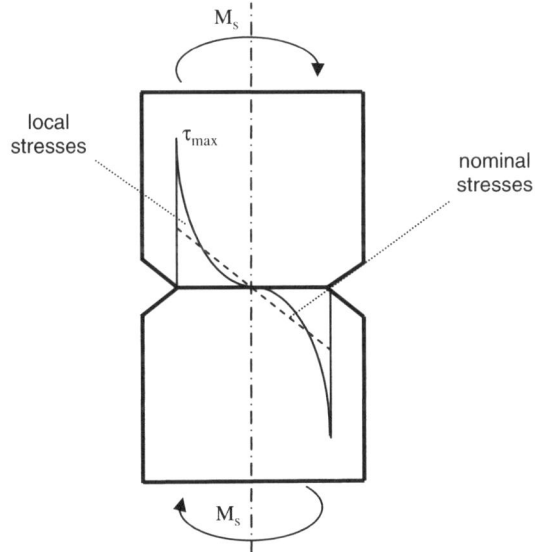

Fig. 2.4. Stress and strain distributions in the element with a sharp notch under torsion

Under tension and bending, stress distributions have been shown for flat and cylindrical elements. In both cases, on the notch root surface plane stress state is observable. In the case of flat elements, inside the material, the plane stress state occurs, and in cylindrical elements the spatial stress state is observed.

Stress and strain distributions in smooth and notched elements have been already analysed in [67, 108, 110, 135, 139, 153, 200, 201, 210, 212, 213]. They were used mainly in non-local methods of fatigue life assessment.

2.2 Theoretical Notch Coefficient

The theoretical notch coefficient is defined as

$$K_t = \frac{\sigma_{xx}^e}{\sigma_{xxn}}. \tag{2.4}$$

According to the Neuber rule, this coefficient can be expressed as the geometric mean from the stress and strain concentration coefficients

$$K_t = \sqrt{K_\sigma K_\varepsilon}, \tag{2.5}$$

which are defined as

$$K_\sigma = \frac{\sigma_{xx}^{e-p}}{\sigma_{xxn}} \tag{2.6}$$

and

$$K_\varepsilon = \frac{\varepsilon_{xx}^{e-p}}{\varepsilon_{xxn}}, \tag{2.7}$$

where ε_{xx}^{e-p} and σ_{xx}^{e-p} are the elastic-plastic strain and the stress in direction of x axis, respectively.

The notch coefficients can be determined in a numerical way, for example with the finite element method, appropriate monograms or suitable formulas, more or less complicated [171, 181, 183, 184 and 185]. There are also other models joining nominal and actual stresses [54], but they are not considered in this book. Up to the yield point, the following relation takes place

$$K_t = K_\sigma = K_\varepsilon. \tag{2.8}$$

For greater stresses, the known relation is valid [20, 24]

$$K_\sigma \leq K_t \leq K_\varepsilon, \quad (2.9)$$

(see Fig. 2.5).

Xiao and Yamada [246] point that the theoretical notch coefficient K_t for welded joints can be determined as a product of the weld geometry action K_w and the influence of structure change in the weld K_s, which can be expressed as

$$K_t = K_w K_s. \quad (2.10)$$

Influence of K_s changes in a welded joint was considered by Chen et al. in [28], and Cheng et al. [29], who tested specimens made of 1Cr–18Ni–9Ti steel under pure tension, pure torsion, and non-proportional tension with torsion. In the specimens tested there was no change of geometry in the joint, i.e. $K_w = 1$. The result scatters for welded joints were greater than those for the native metal. Under pure tension–compression and pure torsion, the fatigue strength of the weld material was less than that of the native material. Such a change was not observed under non-proportional tension with torsion. Thus, the coefficients K_s, including structure changes usually are not separately calculated, and it is assumed that

$$K_t = K_w. \quad (2.11)$$

Whereas, the influence of the K_s coefficient is taken into account during the determination of fatigue notch performance coefficient. The theoretical notch performance coefficient K_t can be used for transformation of stress

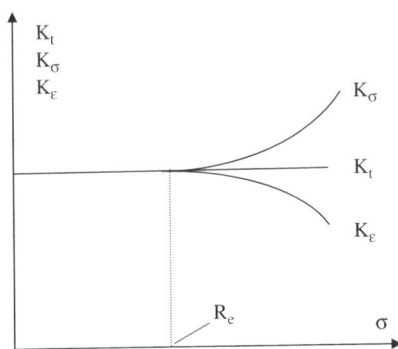

Fig. 2.5. Relation between theoretical notch coefficients and stress value

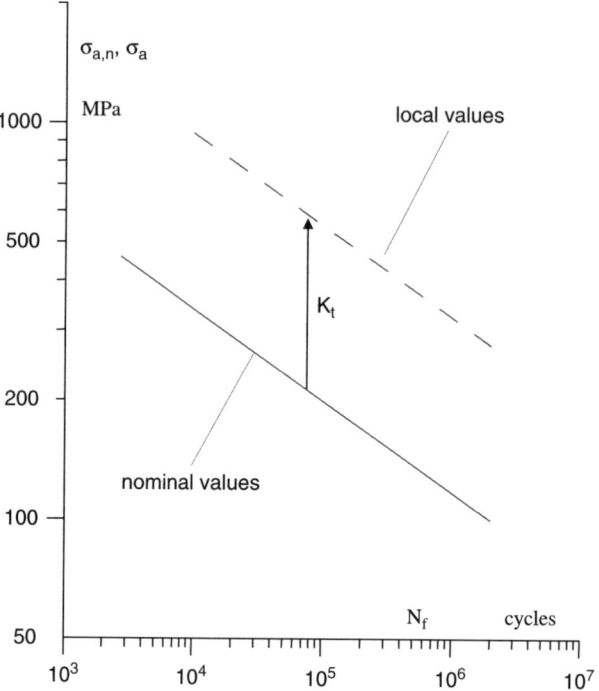

Fig. 2.6. Transformation of stress amplitudes from nominal to local system

amplitude values from nominal to local system, shown in Fig. (2.6) according to (2.4).

2.3 The Fatigue Notch Coefficient

The fatigue notch coefficient K_f [37, 219] is determined by comparison of stresses in smooth, σ_{sm} and notched σ_{not} elements

$$K_f = \frac{\sigma_{sm}}{\sigma_{not}}. \tag{2.12}$$

Interpretation of the coefficient K_f is shown in Fig. 2.7 [1, 2, 3, 119, 121, 135, 189, 190].

The fatigue notch coefficient is usually determined for 10^6 cycles, i.e. [97]

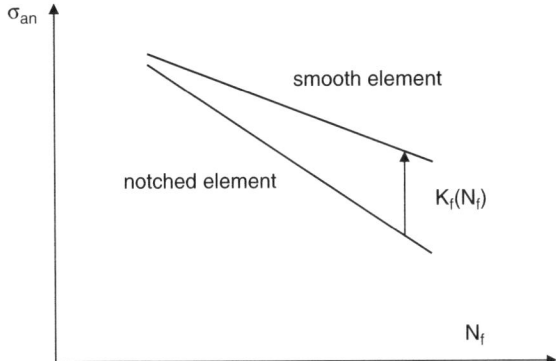

Fig. 2.7. Comparison of nominal stresses σ_{an} for smooth and notched elements

$$K_f = \frac{\sigma_{sm}(10^6 \text{ cycles})}{\sigma_{not}(10^6 \text{ cycles})}. \tag{2.13}$$

From Fig. 2.8 it appears that the fatigue notch coefficient increases as a number of cycles rises. Generally speaking, it can be stated that it is dependent on a number of cycles [20, 43, 197]

$$K_f = \frac{\sigma_{sm}(N_f)}{\sigma_{not}(N_f)}. \tag{2.14}$$

Let us derive a relationship [112]

$$K_f(N_f) = \left(\frac{N_f}{10^3}\right)^{\frac{\log[K_f(10^6)]}{3}}. \tag{2.15}$$

According to this relationship, on the assumption that $N_f = 10^3$ cycles, the fatigue notch coefficient $K_f = 1$. However, it is usually given as a constant value for $N_f = 10^6$ cycles (2.13) (see Fig. 2.8). It can be written as

$$K_f(N_f) = K_f(10^6)\left(\frac{N_f}{10^6}\right)^{\frac{\log[K_f(10^6)]}{3}}. \tag{2.16}$$

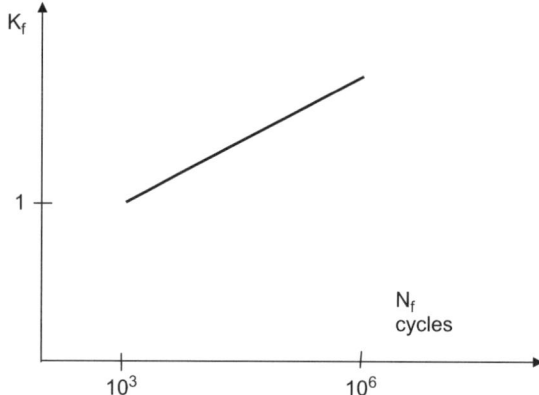

Fig. 2.8. Fatigue notch coefficient K_f versus number of cycles N_f

There are many models joining theoretical and fatigue notch coefficients in the following general form

$$K_f = f(K_t). \tag{2.17}$$

Many papers, among these [175, 247] present relationships among different forms of (2.17) for various materials, types of notches and plastic strains occurring in the notch bottom.

2.4 The Fictitious Radius of the Welding Notch

For determination of notch coefficients the fictitious (conventional) radius in the notch root is applied. Determination of this radius results from stress averaging according to the Neuber's proposal [174]. It is assumed that the crack initiation is controlled by stress in the notch root averaged in a small volume of the material, in the point of the maximum stress occurrence. A suitable material parameter is a substitute microstructural length ρ^*. Stresses in the notch root must be averaged in the interval ρ^* in the direction normal to the surface along this length normal to the notch surface. Taking into account an actual radius in the notch root and the coefficient of multiaxiality s, the expression for the fictitious radius in the notch root is obtained

$$\rho_f = \rho + s\,\rho^*. \tag{2.18a}$$

2.4 The Fictitious Radius of the Welding Notch

For the weld, the worst case can be assumed, i.e. the radius $\rho = 0$, which corresponds to the crack. Then the calculated fictitious radius is expressed as

$$\rho_f = s\,\rho^*. \qquad (2.18b)$$

If the radius is known, it is possible to calculate the notch coefficient, components of the local stress tensor and the corresponding strains. The fictitious radius also depends on geometry of the specimen and a loading mode [159, 173, 174, 202, 208] – they should be taken into account under biaxial bending and torsion (see Table 2.1).

The fictitious notch coefficient ρ_f depends on the actual notch coefficient ρ, the substitute microstructural length ρ^* and the coefficient of multiaxiality s from Table 2.1 (according to the Neuber's proposal) resulting from the stress state multiaxiality in the notch root.

In [229, 230] and some other papers, the authors proposed to determine the substitute microstructural length * according to the following equation

$$\rho^* = \frac{\rho}{s}\left[\frac{(K_t - 1)^2}{(K_f - 1)^2} - 1\right] \qquad (2.19)$$

(see Fig. 2.9).

As it was said above, the zero notch radius, $\rho = 0$, is often assumed with $\rho^* = 0.4$ mm for welded steels, $\rho^* = 0.1$ mm for aluminium alloys [173, 174, 202, 208] and s = 2.5 for plane specimens, when the Huber-Mises-Hencky criterion is used. Then, the fictitious radius $\rho_f = 1$ mm for welded steels and $\rho_f = 0.25$ mm for aluminium is obtained, on the basis of which the

Table 2.1. Coefficients of multiaxiality s according to Neuber [173, 174]

Loading Specimen Criterion	axial or bending plane	Round	shearing or torsion -
Huber-Mises-Hencky	2.5	$\dfrac{5 - 2\nu + 2\nu^2}{2 - 2\nu + 2\nu^2}$	1
Tresca	2	$\dfrac{2 - \nu}{1 - \nu}$	1
maximum normal stresses	2	2	1
Beltrami	$2 - \nu$	$\dfrac{2 - \nu}{1\ \nu}$	1

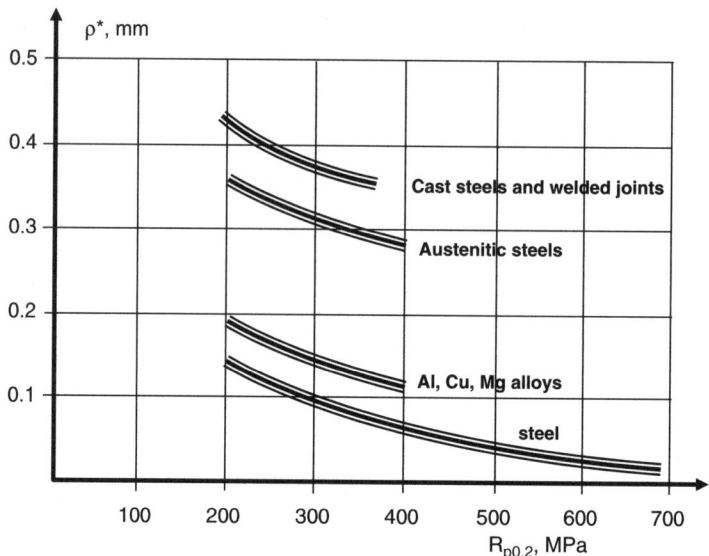

Fig. 2.9. Substitute lengths of microstructure ρ^* for chosen materials and different yield points

determination of the fatigue notch performance coefficient is possible. However, in the case of round specimens subjected to bending, on the assumption that the Poisson's number $\nu = 0.3$, for welded steels the following formula is obtained

$$\rho_{fb} = 0.4 \frac{5 - 2\nu + 2\nu^2}{2 - 2\nu + 2\nu^2} \text{ mm} = 1.16 \text{ mm} \quad (2.20)$$

under torsion the following expression is obtained

$$\rho_{ft} = 0.4 \cdot 1 \text{ mm} = 0.4 \text{ mm}. \quad (2.21)$$

for aluminium subjected to bending the following formula is obtained

$$\rho_{fb} = 0.1 \frac{5 - 2\nu + 2\nu^2}{2 - 2\nu + 2\nu^2} \text{ mm} = 0.29 \text{ mm} \quad (2.22)$$

and for torsion

$$\rho_{ft} = 0.1 \cdot 1 \text{ mm} = 0.1 \text{ mm}. \quad (2.23)$$

2.5 The Notch Coefficient with the Use of the Fictitious Notch Radius

Huther et al. [51] considered fillet joints and analysed influence of the angle of weld face inclination Θ within (30°–55°), and the radius in the notch root ρ within (0.5–3) mm on the fatigue limit. Geometry of such a joint is shown in Fig. 2.10. When the angle rises under stresses determined according to the nominal system, the fatigue limit decreases. The fatigue limit decreases also as the radius in the notch bottom ρ increases.

In [172], influence of the weld face inclination (0°–90°) and the transfer radius (0.6–0.9) mm on the theoretical notch coefficient was considered. Greater notch coefficients are obtained for smaller notch radii. For angles (45°–75°) stabilization of the notch coefficient is observed.

In the weld penetration zone the existence of the fictitious radius in the notch root can be assumed, keeping the same weld face inclination, and then it is possible to determine the theoretical notch coefficient K_t.

Thus, in order to calculate K'_{tb} and K'_{tt} from the fictitious radius in the notch root ρ_f in round specimens, it is necessary to chose separately the fictitious radii for welded elements subjected to bending ($\rho_{fb} = 1.16$ mm) and torsion ($\rho_{ft} = 0.4$ mm), and for elements made of aluminium alloy under bending ($\rho_{fb} = 0.29$ mm) and torsion ($\rho_{ft} = 0.1$ mm) when the constant angle Θ is kept.

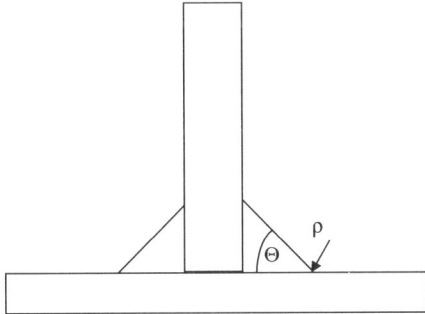

Fig. 2.10. Weld joint with the marked angle of weld face inclination Θ and the radius in the notch root ρ

3 The Stress Model for the Assessment of Fatigue Life Under Uniaxial Loading

3.1 Algorithm for the Assessment of Fatigue Life Under Uniaxial Loading State

Fatigue failure of machine and structure elements caused by service loading often occurs under random stress state. In such a situation, fatigue life is usually calculated with analytic methods or cycle counting methods. The analytic methods use spectral analysis of stochastic processes, and the cycle counting methods are based on numerical algorithms of cycle and half-cycle counting from histories of stress, strain or the energy parameter. The cycle counting methods include schematization of random loading histories, damage accumulation and then fatigue life calculation. Schematization of random histories includes counting of amplitudes and mean values of cycles and half-cycles occurring in the loading history.

In order to define fatigue life under random stress states, determination of the basic fatigue characteristic of the considered material is necessary. It is defined on the basis of cyclic fatigue tests. The basic characteristics for great number of cycles are the stress characteristics in the system $\sigma_a - N_f$, the so-called S–N characteristics. The first characteristic was elaborated by Wöhler [244] in 1860 in a single logarithmic system

$$\log N_f = a + b\sigma_a. \tag{3.1}$$

In 1910, Basquin [16] proposed a characteristic that can be written in a double logarithmic system as

$$\log N_f = a + b\log\sigma_a. \tag{3.2}$$

It is necessary to point out that many authors meaning the Basquin characteristic call it the Wöhler curve. In 1914, Stromeyer [231] presented another proposal including the fatigue limit

$$\log N_f = a + b \log(\sigma_a - \sigma_{af}). \tag{3.3}$$

The next proposals were formulated by Corson in 1955 (see [195])

$$N_f = \frac{a}{\sigma_a - \sigma_{af}} \exp[-c(\sigma_a - \sigma_{af})] \tag{3.4}$$

and Bastenaire in 1974 (see [174, 195]

$$N_f = \frac{a}{\sigma_a - \sigma_{af}} \exp\left[-\left(\frac{\sigma_a - \sigma_{af}}{b}\right)^c\right]. \tag{3.5}$$

Other models were discussed in papers by Palmgren [196], Weibull (1949), Stüssi (1955) and Bastenaire (1963) (see [173]), Kohout, Věchet (2001) [81]. However, the most frequently applied is the Basquin model expressed by (3.2), the so-called S–N fatigue curve.

Under uniaxial loading, fatigue life is calculated according to the algorithm shown in Fig. 3.1 and the stress model [17, 18, 94, 95, 127, 143, 144, 145, 148, 160, 164]. A similar algorithm for uniaxial loading has been proposed by Gołoś [43].

Stage 1
The input data for fatigue life calculations are strain $\varepsilon(t)$ or stress $\sigma(t)$ histories, which can be obtained from:

- measurements of actual strains [41] or forces (strain gauges, extensometers, force gauges). Under uniaxial tension and on the assumption of a

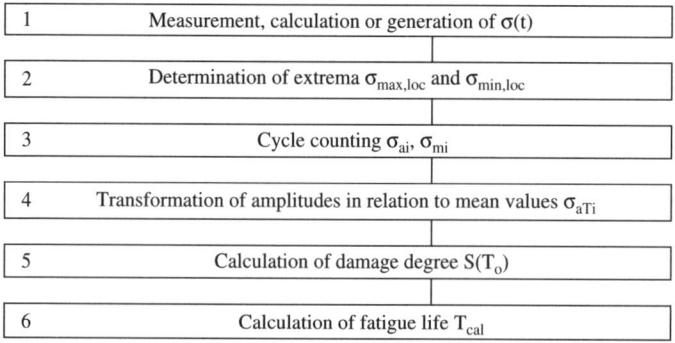

Fig. 3.1. Algorithm for determination of fatigue life under uniaxial random loading

perfectly elastic body, the relationship between the stress and strain histories can be written as

$$\sigma(t) = E\varepsilon(t), \tag{3.6}$$

- previous numerical calculations [85] (FEM – finite element method, BEM – boundary element method, FDM – finite difference method),
- computer generation of random sequences with shaped probabilistic characteristics corresponding to service conditions or the predicted states. Standard programs elaborated in some research centers can be used for this purpose. Some well-known standards are: WASH1 for loading simulation in drilling platforms [49, 205], Broad64 and MMMOD64 [6] (see also [5]) – for drilling platforms, too, CARLOS [206] for car wheel loading, wind load [25]. Other possibilities of generation of signals have been presented, among others, in [26, 52, 85, 170, 198, 211, 236, 245, 248].

Stage 2

At this stage, extrema of the stress history are defined. Under random history, values of successive extrema are determined. This process includes observation of the derivative from the history and search of its monotonic changes, see Fig. 3.2. Figure 3.2 shows some exemplary determined local minima (2, 4, 6, 8) and extrema (1, 3, 5, 7, 9) in a course fragment.

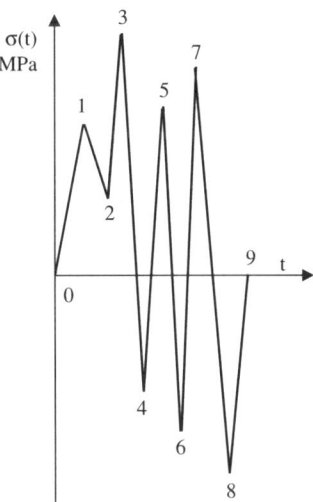

Fig. 3.2. Determination of local extrema

Stage 3

In the case of random histories with wide frequency bands, several cycle counting methods can be applied [80, 238]. The full cycle method in which half-cycles are not included, and the obtained life is usually overestimated. However, with the use of three other methods, i.e. the methods of range pairs, hysteresis loop and rain flow [10, 34, 35], the cycles, half-cycles and their mean values can be determined. These three methods usually give similar results. In practice, the rain flow method (so-called envelope method) is most often applied. Its scheme is shown in Fig. 3.3. Envelopes are drawn from each local extremum (maximum or minimum). If the envelope has its beginning at the local minimum, it ends at the local maximum located opposite the local minimum, the value of which is lower than the initial minimum. The envelope beginning at the local minimum 0 ends at the local maximum 3 opposite the local minimum 4. The same procedure is applied when the envelope begins at the local maximum. Then it ends at the local minimum opposite the local maximum, the value of which is higher than the initial maximum. The envelope beginning at the local maximum 1 ends at the point 2 located opposite the local maximum 3. Half-cycles should be isolated from the determined cycles. The half-cycle of the largest span is included between the global maximum and global minimum (3 and 8). If the local minimum occurs as the first local minimum, the half-cycle is determined between this local minimum and the global maximum (0 and 3), and between this global maximum and the preceding local minimum (3 and 8). One more half-cycle is obtained between the global minimum and the last local maximum (8 and 9). The same procedure is applied when the beginning of the cycle is in the local maximum.

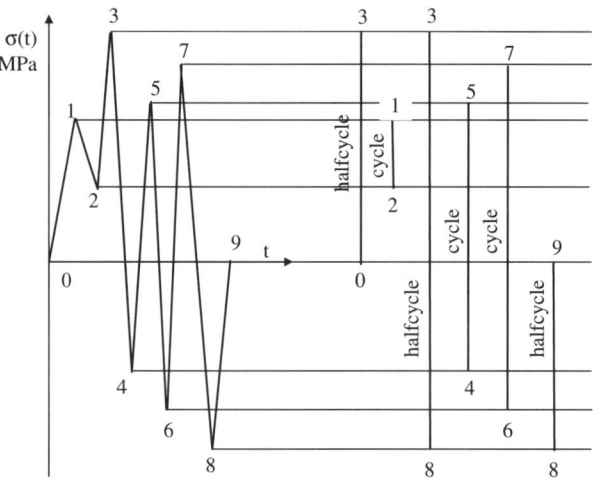

Fig. 3.3. Cycle counting with the rain-flow method [7, 34, 43, 80]

As it was mentioned above, the rain flow method, so-called envelope method, allows to define both cycles and half-cycles, which are determined by suitable envelopes (see Fig. 3.3). This method has been programmed and cycles are counted by the computer program. The amplitude σ_{ai} and the mean value σ_{mi} of a cycle or a half-cycle are determined each time.

Stage 4
At this stage, transformation of cycle amplitudes σ_{ai} takes place in relation to the occurring mean values σ_{mi} according to the general equation for the transformed amplitude, analysed in some previous papers [74, 75, 76, 113, 146, 147],

$$\sigma_{aTi} = f(\sigma_{ai}, \sigma_{mi}). \qquad (3.7)$$

There are many models that take into account the influence of mean values. In this paper, the above transformations have not been widely presented because of the fact that in welded joints high residual stresses are often observed and then probable loading with the mean value occurring while a cycle do not influence the fatigue life.

Stage 5
There are many hypotheses of fatigue damage accumulation (stage 4) [235] (linear and nonlinear). The linear hypotheses proposed by Palmgren–Miner [169, 196], Haibach [47] and Serensen-Kogayev [209], Corten-Dolan [30], Liu-Zenner [99] are most frequently applied.

Damages can be accumulated according to the Palmgren–Miner hypothesis [169, 196], including amplitudes below the fatigue limit and the coefficient $a \leq 1$

$$S_{PM}(T_o) = \begin{cases} \sum_{i=1}^{j} \dfrac{n_i}{N_o \left(\dfrac{\sigma_{af}}{\sigma_{ai}} \right)^m} & \text{for } \sigma_{ai} \geq a \cdot \sigma_{af} \\ 0 & \text{for } \sigma_{ai} < a \cdot \sigma_{af} \end{cases}, \qquad (3.8)$$

Haibach hypothesis [46]

$$S_H(T_o) = \begin{cases} \sum_{i=1}^{j} \dfrac{n_i}{N_o \left(\dfrac{\sigma_{af}}{\sigma_{ai}}\right)^m} & \text{for } \sigma_{ai} \geq \sigma_{af} \\ \sum_{i=j}^{k} \dfrac{n_i}{N_o \left(\dfrac{\sigma_{af}}{\sigma_{ai}}\right)^{2m-p}} & \text{for } \sigma_{ai} < \sigma_{af} \end{cases}, \quad (3.9)$$

where [223]:
 p =1 for steels and aluminium alloys,
 p =2 for casts and sintered steels,

Serensen-Kogayev hypothesis [209]

$$S_{SK}(T_o) = \begin{cases} \sum_{i=1}^{j} \dfrac{n_i}{bN_o \left(\dfrac{\sigma_{af}}{\sigma_{ai}}\right)^m} & \text{for } \sigma_{ai} \geq a \cdot \sigma_{af} \\ 0 & \text{for } \sigma_{ai} < a \cdot \sigma_{af} \end{cases}, \quad (3.10)$$

where:

$$b = \dfrac{\sum_{i=1}^{k} \sigma_{ai} t_i - a \cdot \sigma_{af}}{\sigma_{a\,max} - a \cdot \sigma_{af}} \quad \text{for } b > 0.1, \quad (3.11)$$

is the Serensen-Kogayev coefficient, connected with a history character, and

$$t_i = \dfrac{n_i}{\sum_{i=1}^{k} n_i} \quad (3.12)$$

is frequency of occurrence of particular levels σ_{ai} in observation time T_0, and σ_{af} is the general fatigue limit. The relationship (3.11) is valid if the following condition is satisfied $\dfrac{\sigma_{a\,max}}{\sigma_{af}} > 1$ and $\dfrac{1}{\sigma_{a\,max}} \sum_{i=1}^{k} \sigma_{ai} t_i > 0.5$.

Corten-Dolan hypothesis [30]

$$S_{CD}(T_o) = \begin{cases} \sum_{i=1}^{j} \dfrac{n_i}{N_1 \left(\dfrac{\sigma_{a\,max}}{\sigma_{ai}}\right)^{m'}} & \text{for } \sigma_{ai} \geq \sigma_{af} \\ 0 & \text{for } \sigma_{ai} < \sigma_{af} \end{cases}, \quad (3.13)$$

where

$$m' = (0.8\text{–}0.9)\,m, \quad N_1 = N_o \left(\dfrac{\sigma_{af}}{\sigma_{a\,max}}\right)^{m} \quad (3.14)$$

Liu-Zenner hypothesis [99]

$$S_{LZ}(T_o) = \begin{cases} \sum_{i=1}^{j} \dfrac{n_i}{N_1 \left(\dfrac{\sigma_{a\,max}}{\sigma_{ai}}\right)^{m'}} & \text{for } \sigma_{ai} \geq a \cdot \sigma_{af} \\ 0 & \text{for } \sigma_{ai} < a \cdot \sigma_{af} \end{cases}, \quad (3.15)$$

where

$$m' = \dfrac{m + m_i}{2}, \quad N_1 = N_o \left(\dfrac{\sigma_{af}}{\sigma_{a\,max}}\right)^{m}, \quad (3.16)$$

and m_i is a slope of the fatigue curve S–N for fatigue crack initiation.

It is important to note that the models (3.12) and (3.15) act in a similar way as the Liu-Zenner model, however the Liu-Zenner model explains a new slope of the fatigue curve.

A model similar to the Serensen-Kogayev proposal was formulated for alloys of non-ferrous metals [58]. From calculations, b less or greater than 1 is obtained, depending on the mean-square weighed amplitude [107, 157]

$$S_{K\text{Ł}}(T_o) = \begin{cases} \dfrac{\sum_{i=1}^{j} n_i}{b' N_o \left(\dfrac{\sigma_{af}}{\sigma_{ai}}\right)^m} & \text{for } \sigma_{ai} \geq a \cdot \sigma_{af} \\ 0 & \text{for } \sigma_{ai} < a \cdot \sigma_{af} \end{cases} \qquad (3.17)$$

where a coefficient including a damage degree (D≠1) is determined. This coefficient characterizes the history and takes the form

$$b' = \frac{\sigma(N_f)}{\sigma_{aw}}, \qquad (3.18)$$

where

σ_{aw} – stress amplitude for the given number of cycles, expressed by the following equation:

$$\sigma_{aw} = \left[\frac{\sum_i \sigma_{ai}^2}{\sum_i n_i}\right]^{\frac{1}{2}} \quad \text{for } \sigma_{ai} > a\sigma_{af}, \qquad (3.19)$$

σ_{ai} – stress amplitude,
n_i – a number of stress cycles with amplitude σ_{ai}.

where: σ_{aw} – stress amplitude for the given number of cycles.
  In the discussed calculations, it was assumed that a number of cycles N_f was equal to N_0, so:

$$\sigma(N_f) = \sigma_{af}. \qquad (3.20)$$

Thus, when

$\sigma_{aw} > \sigma(N_f)$ then $b' < 1$
$\sigma_{aw} = \sigma(N_f)$ then $b' = 1$.
$\sigma_{aw} < \sigma(N_f)$ then $b' > 1$

  The damage degree changes depending on a level and a number of stress amplitudes. It decreases as the amplitudes increase. Figure 3.4 shows a

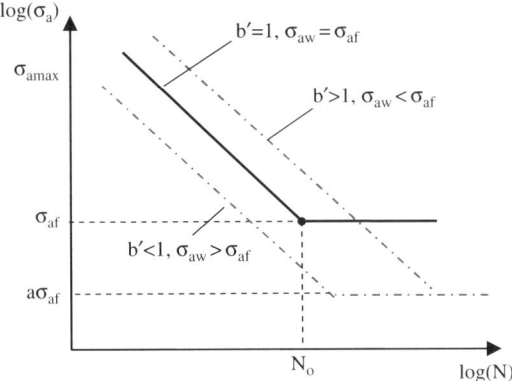

Fig. 3.4. Changes of the coefficient b' for $\sigma(N_f) = \sigma_{af}$

scheme of changes of the coefficient b' depending on the weighed amplitudes for the S–N curve.

The hypotheses (3.8, 3.9, 3.10, 3.13, 3.15, 3.17) can be written as one expression:

$$S(T_o) = \begin{cases} \sum_{i=1}^{j} \dfrac{n_i}{b * N * (\sigma_{af} / \sigma_{ai})^m} & \text{dla } \sigma_{ai} \geq a\sigma_{af} \\ h \sum_{i=j+1}^{k} \dfrac{n_i}{N * (\sigma_{af} / \sigma_{ai})^{(2m-p)}} & \text{dla } \sigma_{ai} < a\sigma_{af} \end{cases}, \quad (3.21)$$

where:
$S(T_o)$ – material damage degree at time T_o according to (3.8, 3.9, 3.10, 3.13, 3.15) or (3.17),
n_i – a number of cycles with amplitudes σ_{ai} in T_o,
T_o – observation time (for analysis of loadings with variable amplitudes a number of cycles in one block, N_{bloc} is assumed),
m – exponent of the S–N fatigue curve,
m' – modified slope coefficient for the S–N fatigue curve for Corten-Dolan (3.13) and Liu-Zenner (3.17) hypotheses, in another case m'=m,
N_o – a number of cycles corresponding to the fatigue limit σ_{af},
$N^* = N_1$ for Corten – Dolan (3.13) and Liu – Zenner (3.17), in another case $N^* = N_1$,
k – a number of class intervals of the amplitude histogram (j < k),
a – coefficient allowing to include amplitudes below σ_{af} in the damage accumulation process, (for Haibach (3.9) and Corten-Dolan (3.13) a = 1),

b* – coefficient including history character; for Serensen-Kogayev (3.10) b* = b, for Kardas-Łagoda [57] (3.17) b*= b, in other cases b* = 1,
p – coefficient modifying the fatigue curve according to Haibach for amplitudes below the fatigue limit,
h – coefficient for the Haibach hypothesis (3.9) h =1 (for other hypotheses h = 0).

General forms are shown in Fig. 3.5, and method of damage accumulation according to the Palmgren–Miner rule is presented in Fig. 3.6.

As it was stated before, the assumption of linear summation of fatigue damage with modifications was proved many times during experiments

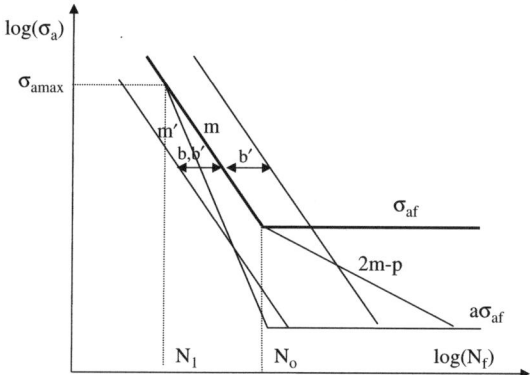

Fig. 3.5. Original Basquin fatigue curve and its modifications for fatigue damage accumulation

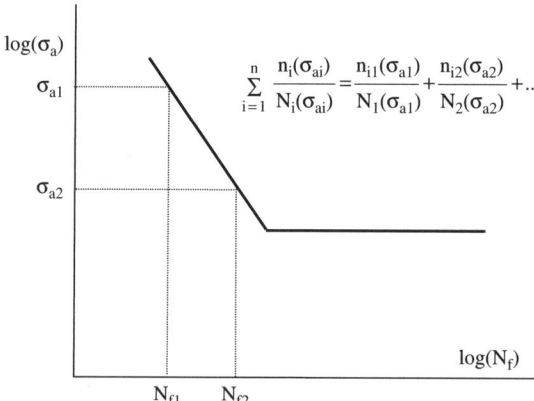

Fig. 3.6. A way of fatigue damage accumulation

under uniaxial loading being a stationary stochastic process of normal probability distribution.

Stage 6

After determination of a damage degree during observation time T_o according to a general form (3.21), fatigue life is determined

$$T_{cal} = \frac{T_o}{S(T_o)}. \qquad (3.22)$$

After determination of the damage degree $S(N_{block})$ for a number of cycles N_{block} in a loading block according to the general formula (3.21), fatigue life is calculated according to the following equation

$$N_{cal} = \frac{N_{block}}{S(N_{block})}. \qquad (3.23)$$

3.2 Statistic Evaluation

From references, for example from [96], it appears that for fatigue tests large scatters are typical. There are scatters of life under a given loading or scatters of loading (stresses, strains, the strain energy density parameter) under the given life.

From the paper by Lahti et al. [96] it appears that in the case of life of welded joints test results are included in the scatter band with the coefficient about 4 with probability 95%. It is defined as life scatters, i.e.

$$T_N = N_{cal}/N_{exp}, \qquad (3.24)$$

or inverse of (3.24)

$$T'_N = 1/T_N = N_{exp}/N_{cal}. \qquad (3.25)$$

Ratios (3.22) and (3.25) can be called the scatter band with coefficient T_N. This scatter band varies depending on materials, loading level and mode, and it is included within the range from 1.5 for a low number of cycles, steel and notched specimens to 5 for a level close to the fatigue limit, cast iron or welded joints. The least scatters are obtained for tension where all the section area is equally subject to cracking (Fig. 3.7). A little greater

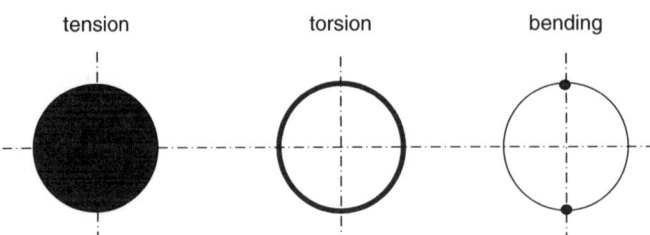

Fig. 3.7. The most loaded parts of the section under different simple loadings

scatters are often obtained under torsion (in the case of notches they are much greater), where the greatest stresses occur at the perimeter. The greatest scatters can be observed for bending, where only extreme fibres are loaded to a highest degree.

In [90, 91, 232], the error of life determination was related to the experimental life according to the following formula

$$E = \frac{|N_{cal} - N_{exp}|}{N_{exp}}. \tag{3.26}$$

In [12] a standard deviation of logarithms of the calculated lives related to the experimental ones is presented

$$s_N = \sqrt{\frac{\sum_{i=1}^{n}(\log N_{cal} - \log N_{exp})^2}{n-1}}. \tag{3.27}$$

In [21], a scatter of the calculation results related to the experimental results is considered

$$T_N = \frac{T_{N10\%}}{T_{N90\%}}, \tag{3.28}$$

where T_N is defined by (3.28), for 10% and 90% of probability of damage, respectively.

Other authors analyse the stress scatters.

Bellet et al. [19] compare calculated and experimental lives trying to assess efficiency of the models according to the following formula

$$E = \frac{\sigma_{exp}}{\sigma_{cal}}. \tag{3.29}$$

Another possibility of stress comparing is proposed by Sonsino and co-authors in many papers, for example [36, 219, 220, 222, 239] who apply the following equation:

$$T_\sigma = \frac{\sigma(P = 10\%)}{\sigma(P = 90\%)} \text{ for } N_f = \text{const.} \tag{3.30}$$

Such a formula includes only 80% calculating points.
Equation (3.30) for life can be written as:

$$T_N = T_\sigma^m, \tag{3.31}$$

and in this case T_N is defined according to (3.31) as:

$$T_N = \frac{N_f(P = 10\%)}{N_f(P = 90\%)}. \tag{3.32}$$

There are also models that include scatters of damage degrees analysis (see [27]), which is similar to (3.23)

$$E = \frac{|D_{cal} - D_{exp}|}{D_{exp}}. \tag{3.33}$$

During the analysis of life scatters the life ratios according to (3.24) or (3.25), or logarithms of life are usually used, according to

$$E = \log \frac{N_{exp}}{N_{cal}} \tag{3.34}$$

(see [11]).
The mean value of the considered quantity can be defined as

$$\overline{E} = \frac{1}{n}\sum_{i=1}^{n} E_i, \tag{3.35}$$

and the mean error of the mean value can be defined as

$$SE = \frac{s}{\sqrt{n}}, \tag{3.36}$$

where
n – a number of measurements,
s – mean standard deviation.

For determination of the variance the following formula should used

$$s^2 = \frac{1}{n-1} \sum_{i=1}^{n} (E_i - \overline{E})^2, \tag{3.37}$$

and the standard deviation should be determined from variance (3.34)

$$s = \sqrt{s^2}. \tag{3.38}$$

In the case of material fatigue, the significance level is usually assumed at the minimum level α = 5% or 10%, sometimes even 20%. Thus, the mean value should be included within the range

$$-t_{(n-1),\alpha/2}(SE) \leq \overline{E} \leq t_{(n-1),\alpha/2}(SE) \tag{3.39}$$

or

$$-t_{(n-1),\alpha/2}(s) \leq \overline{E} \leq t_{(n-1),\alpha/2}(s), \tag{3.40}$$

where $t_{(n-1),\alpha/2}$ – constant from the t-Student's distribution for the mean value error SE – (3.36) or the population error s – (3.38).

Constant $t_{(n-1),\alpha/2}$ from the t-Student's distribution is determined for a half of the significance level α/2 because of section of the normal distribution edges (see Fig. 3.8).

The mean scatter is determined from the following relationship

$$\overline{T}_N = 10^{\overline{E}} \tag{3.41}$$

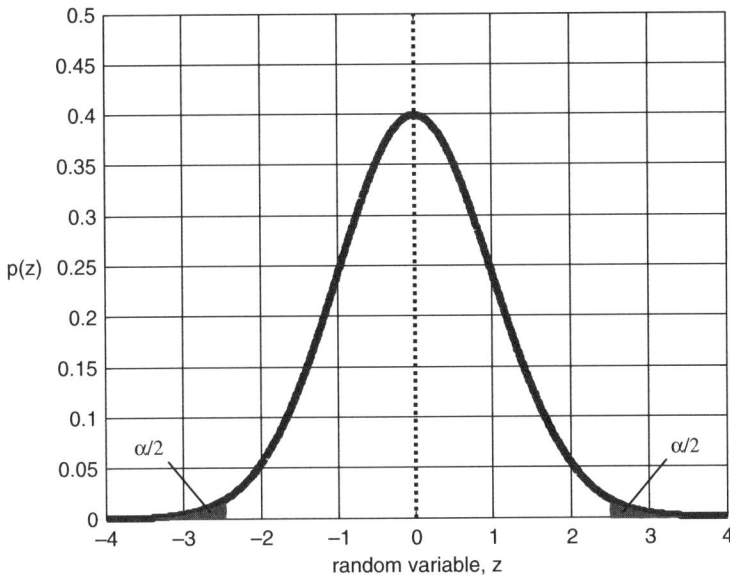

Fig. 3.8. Normal distribution with sections at significance level α

in the scatter band with the scatter coefficient T_N expressed as

$$T_N = 10^{t_{n-1}\alpha/2 \cdot s}. \quad (3.42)$$

For the significance level α/2 = 2.5% and n = 60 (often 20–30 measurements) the scatter band for all population is obtained. It is equal to two standard deviations (2s) (or maximum 2.2s for 20 measurements), which is often applied in tests [166] and corresponds to the scatter band with the coefficient 3. The scatters 3s correspond to a large significance level tending to zero but they are not applied in fatigue tests [9].

4 The Energy Model of Fatigue Life Assessment

For the complex loading state the energy model is being proposed. The model is based on the strain energy density parameter (SEDP) and analyses changes of stress (normal and shear) and strain (normal and shear) in the critical plane. It also distinguishes tension and compression.

4.1 The Energy Parameter Under Uniaxial Loading

The change of strain energy density, widely used in theory of plasticity, is also proposed as a parameter of the multiaxial fatigue analysis. Suitability of this parameter for description of fatigue processes seems to be promising, especially while formulation of thermal-elastic-plastic models of strain in the materials subjected to random thermomechanical loading. The models do not include a division of strain energy density into elastic and plastic parts, like in case of the parameters proposed by Smith–Watson–Topper (SWT) [214], Hoffman and Seeger [48], Bergman and Seeger [22]. In the elastic range, energy can be calculated from

$$W = \tfrac{1}{2}\sigma\varepsilon. \tag{4.1}$$

In the time domain, energy density can be expressed as

$$W(t) = \tfrac{1}{2}\sigma(t)\varepsilon(t). \tag{4.2}$$

When the equation is connected with the damage parameter

$$P_{SWT} = \sqrt{\sigma_{max}\varepsilon_a E}, \tag{4.3}$$

where

$$\sigma_{max} = \sigma_m + \sigma_a, \tag{4.4}$$

and when the mean value of stress is equal to zero

$$\sigma_{max} = \sigma_a, \qquad (4.5)$$

the strain energy density takes the form

$$W = \frac{P_{SWT}^2}{2E}. \qquad (4.6)$$

It should be pointed that parameter P_{SWT} has a dimension of stress. However, it is expressed as energy per a volume unit, i.e. MJ/m^3.

Further modification of the considered parameter can be written as [22]

$$P = \sqrt{(k\sigma_m + \sigma_a)\varepsilon_a E} \qquad (4.7)$$

or for shear stresses and shear strains [48]

$$P_\tau = \sqrt{\tau_a \gamma_{max} G}. \qquad (4.8)$$

In order to distinguish tension and compression in a fatigue cycle, functions sgn[$\varepsilon(t)$] and sgn[$\sigma(t)$] should be substituted to (4.2):

$$W(t) = \tfrac{1}{4}\sigma(t)\varepsilon(t)\,\text{sgn}[\varepsilon(t)] + \tfrac{1}{4}\sigma(t)\varepsilon(t)\,\text{sgn}[\sigma(t)] = \\ = \tfrac{1}{4}\sigma(t)\varepsilon(t)\{\text{sgn}[\varepsilon(t)] + \text{sgn}[\sigma(t)]\} = \tfrac{1}{2}\sigma(t)\varepsilon(t)\,\tfrac{\text{sgn}[\varepsilon(t)] + \text{sgn}[\sigma(t)]}{2}. \qquad (4.9)$$

A two-argument logical function is sensitive to the signs of variables, and it is defined as

$$\text{sgn}(x,y) = \tfrac{\text{sgn}(x)+\text{sgn}(y)}{2} = \begin{cases} 1 & \text{when } \text{sgn}(x)=\text{sgn}(y)=1 \\ 0.5 & \text{when } (x=0 \text{ and } \text{sgn}(y)=1) \text{ or } (y=0 \text{ and } \text{sgn}(x)=1) \\ 0 & \text{when } \text{sgn}(x)=-\text{sgn}(y) \\ -0.5 & \text{when } (x=0 \text{ and } \text{sgn}(y)=-1) \text{ or } (y=0 \text{ and } \text{sgn}(x)=-1) \\ -1 & \text{when } \text{sgn}(x)=\text{sgn}(y)=-1 \end{cases},$$

$$(4.10)$$

where

$$\text{sgn}(x) = \begin{cases} 1 & \text{when } x > 0 \\ 0 & \text{when } x = 0 \\ -1 & \text{when } x < 0 \end{cases}. \qquad (4.11)$$

After the substitution sgn(x,y) to (4.9) the following formula is obtained

$$W(t) = \tfrac{1}{2}\sigma(t)\varepsilon(t)\,\mathrm{sgn}[\sigma(t),\varepsilon(t)]. \tag{4.12}$$

Equation (4.12) expresses positive and negative values of the strain energy density parameter in a fatigue cycle and it allows to separate energy (work) under tension from energy (work) under compression. If the parameter is positive, it means that the material is subjected to tension. If the parameter is negative, the material is subjected to compression with energy equal to this parameter for the absolute value. Equation (4.12) has another advantage: a course of the strain energy density parameter has the zero mean value, and the cyclic stress and strain have also the expected zero value, i.e. R = −1. Moreover, when stress or strain reaches zero, (4.9) is equal to zero, so sgn(x, y) = 0.5 does not occur. Thus, (4.10) can be written in the reduced form

$$\mathrm{sgn}(x,y) = \frac{\mathrm{sgn}(x)+\mathrm{sgn}(y)}{2} = \begin{cases} 1 & \text{when } \mathrm{sgn}(x)=\mathrm{sgn}(y)=1 \\ 0 & \text{when } \mathrm{sgn}(x)=-\mathrm{sgn}(y) \\ -1 & \text{when } \mathrm{sgn}(x)=\mathrm{sgn}(y)=-1. \end{cases} \tag{4.13}$$

Figure 4.1a shows a course of the energy parameter according to (4.6), and the strain energy density parameter, taking into account signs of both stress and strain for an elastic body. If the signs of stresses and strains are not taken into account (Fig. 4.1b), a number of cycles with small ranges of energy parameters (Fig. 4.1b) is doubled and a non-zero mean value is obtained. This model is valid for R = −1.

If cyclic stresses and strains reach their maximum values, σ_a and ε_a, then the amplitude of maximum strain energy density parameter – according to (4.9) – is

$$W_a = 0.5\sigma_a \varepsilon_a. \tag{4.14}$$

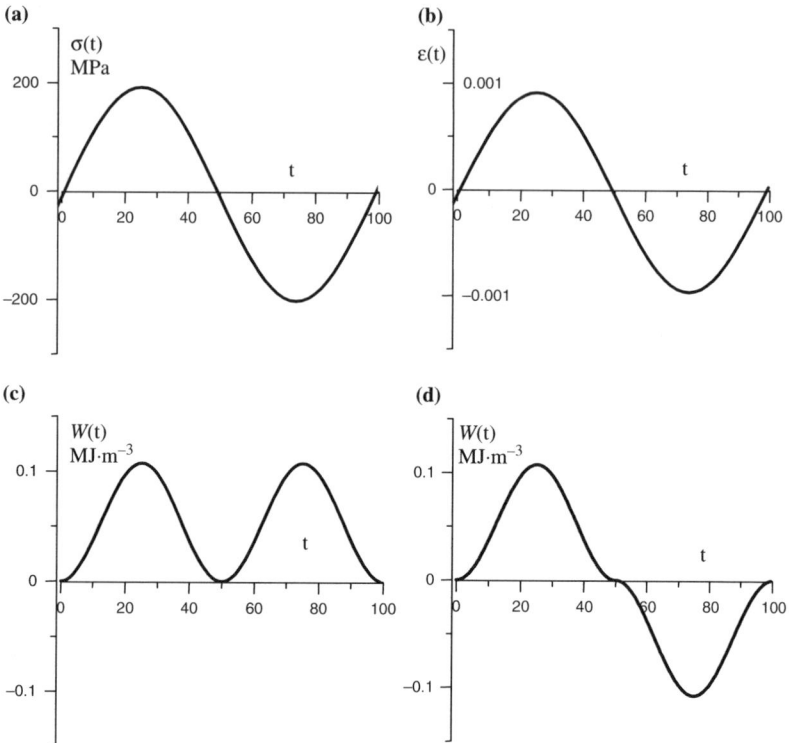

Fig. 4.1. Histories of cycle of stress, strain and strain energy density parameter – (4.2) and strain energy parameter including signs of stress and strain – (4.13)

Assuming - according to (4.9) – that W is the fatigue damage parameter, the standard characteristics of cyclic fatigue can be rescaled and it is possible to obtain a new characteristic $(W_a - N_f)$ for low- and high-cycle fatigue. Under high-cycle fatigue, when the curve $(\sigma_a - N_f)$ is applied, the axis σ_a should be replaced by W_a, where

$$W_a = \frac{\sigma_a^2}{2E}. \tag{4.15}$$

Using the Manson-Coffin-Basquin equation

$$\varepsilon_a = \varepsilon_a^e + \varepsilon_a^p = \frac{\sigma'_f}{E}(2N_f)^b + \varepsilon'_f (2N_f)^c \tag{4.16}$$

4.1 The Energy Parameter Under Uniaxial Loading

and (4.14), we obtain the strain energy density parameter

$$W_a = \frac{\sigma_a}{2}\left[\frac{\sigma'_f}{E}(2N_f)^b + \varepsilon'_f(2N_f)^c\right]. \tag{4.17}$$

Equation (4.17) gives a new description of fatigue history, the curve $(W_a - N_f)$, i.e.

$$W_a = \frac{(\sigma'_f)^2}{2E}(2N_f)^{2b} + 0.5\varepsilon'_f\,\sigma'_f\,(2N_f)^{b+c}. \tag{4.18}$$

Under high-cycle fatigue, (4.18) reduces to the following simple form

$$W_a = \frac{(\sigma'_f)^2}{2E}(2N_f)^{2b} = \frac{(\sigma'_f)^2}{2E}(2N_f)^{b'}. \tag{4.19}$$

After finding the logarithm, further reduction takes place and the following formula is obtained

$$\log N_f = A' - m' \log W_a, \tag{4.20}$$

where:

$$A' = -\frac{1}{b'}\left[\log\frac{(\sigma'_f)^2}{2E} + b'\log 2\right], \tag{4.21}$$

$$m' = -\frac{1}{b'}, \tag{4.22}$$

$$b' = 2b. \tag{4.23}$$

Finally, the following expression is obtained

$$m' = \frac{1}{2b} = \frac{m}{2}. \tag{4.24}$$

Under high-cycle fatigue, control of stress and strain are very close, especially in the case of cyclically stable materials. The constants A' and m' in (4.20) are determined from the fatigue curve S–N, by simple rescaling (4.18), determined on the basis of tests under controlled strain.

According to (4.15), replacing the stress amplitude by fatigue strength σ_{af} for a given fatigue life, the strain energy density parameter at the fatigue limit level is obtained

$$W_{af} = \frac{\sigma_{af}^2}{2E}. \tag{4.25}$$

Introducing the characteristic for shear strains

$$\gamma_a = \frac{\tau'_f}{G}(2N_f)^{b_\tau} + \gamma'_f (2N_f)^{c_\tau} \tag{4.26}$$

a new strain characteristics expressed by the parameter of shear strain energy density is obtained

$$W_a = \frac{(\tau'_f)^2}{2G} N_f^{2b_\tau} + \frac{1}{2}\tau'_f \gamma'_f N_f^{b_\tau+c_\tau} \tag{4.27}$$

or, in the case of elastic strains

$$\log N_f = A'_\tau - m'_\tau \log W_a, \tag{4.28}$$

Figure 4.2 shows random histories of stress, strain and the parameter of normal strain energy density. From this figure it appears that in the case of the strain energy density parameter and neglecting signs of stresses and strains (Fig. 4.2c) the frequency band extension is obtained, and – in consequence – counting a greater number of cycles with mean values different from zero under the generated zero mean values of stresses and strains (Fig. 4.2a and b). On the other hand, using the strain energy density parameter and including signs of stresses and strains (Fig. 4.2d), it is possible to obtain a history with the zero mean value without extension of the frequency band, like for cyclic loading (Fig. 4.1).

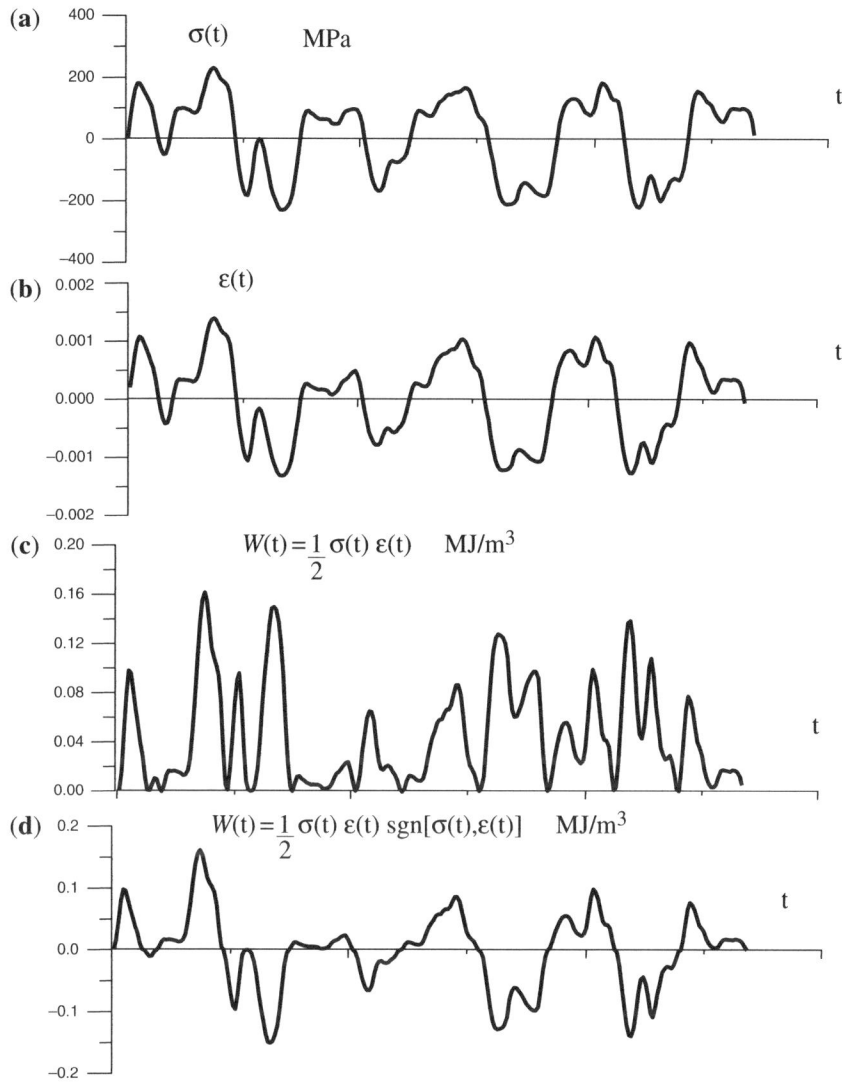

Fig. 4.2. Random history of stress, strain, strain energy density and normal strain energy density parameter

4.2 The Energy Parameter Under Multiaxial Loading

Analogous to the case of uniaxial loading state presented in Sect. 4.1 [115, 120, 123,124, 131, 136, 164, 188], the proposed generalized energy criterion is based on the analysis of stresses and the corresponding strains in the critical plane, taking their signs into account. These criteria were formulated on the basis of previous considerations and papers [15, 82, 104, 111, 115, 117, 123, 129, 133, 134, 135, 137, 191, 192, 193]. The criteria valid under stress concentration, i.e. including the complex stress state under uniaxial loading, are presented below.

4.2.1 The Generalized Criterion of the Parameter of Normal and Shear Strain Energy Density Parameter in the Critical Plane

In the case of the proposed damage parameter under stress concentration the same assumptions as those for smooth elements can be made.

Fatigue cracking is caused by the part of strain energy density corresponding to work of the normal stress $\sigma_\eta(t)$ on the normal strain $\varepsilon_\eta(t)$, i.e. $W_\eta(t)$, and work of the shear stress $\tau_{\eta s}(t)$ on the shear strain $\varepsilon_{\eta s}(t) = 0.5\gamma_{\eta s}(t)$ in direction $\bar{s}$ on the plane with normal $\bar{\eta}$, i.e. $W_{\eta s}(t)$;

1. Direction $\bar{s}$ on the critical plane coincides with the mean direction of maximum shear strain energy density $W_{\eta s max}(t)$;
2. In the limit state, material effort is defined by the maximum value of linear combination of energy parameters $W_\eta(t)$ and $W_{\eta s}(t)$, where energy satisfies the following equation under multiaxial random loading

$$\max_t \{\beta W_{\eta s}(t) + \kappa W_\eta(t)\} = Q \tag{4.29}$$

or

$$\max_t \{W(t)\} = Q, \tag{4.30}$$

where β is the constant for a particular form of (4.29), and κ and Q are material constants determined from sinusoidal simple fatigue tests.

The left sides of (4.29) and (4.30) can be written as maximum at time W(t) and they should be understood as 100% quantile of a random variable W. If the maximum value W(t) exceeds Q, then damage accumulation and failure take place. The random process W(t) can be interpreted as a stochastic process of the material fatigue strength. Positions of unit vectors

$\bar{\eta}$ and $\bar{s}$ are determined with the weight function method, variance method or damage accumulation method [138].

Selection of constants β, κ and Q in (4.29), and the assumed position of the critical plane leads to particular cases of the generalized criterion.

In special forms of the criterion a complex stress state should be assumed on the surface of the notched element. This is also necessary in the case of states under simple loading, such as bending or axial loading (tension-compression).

The equivalent value of the strain energy density parameter is a linear combination of energy density of normal and shear strains. Participation of particular energies in the damage process depends on the coefficients β and κ. In each case the scalar product of vectors is $\bar{\eta} \circ \bar{s} = 0$, where the vectors are defined as:

$$\bar{\eta} = \hat{l}_\eta \bar{i} + \hat{m}_\eta \bar{j} + \hat{n}_\eta \bar{k}, \qquad (4.31)$$

$$\bar{s} = \hat{l}_s \bar{i} + \hat{m}_s \bar{j} + \hat{n}_s \bar{k}. \qquad (4.32)$$

The general criterion (4.29) can be written as

$$\max_t \{\beta W_{\eta s}(t) + \kappa W_\eta(t)\} = W_{af}, \qquad (4.33)$$

From (4.33) the equation for the equivalent strain energy density parameter can be derived

$$W_{eq}(t) = \beta W_{\eta s}(t) + \kappa W_\eta(t), \qquad (4.34)$$

where:

$$W_{\eta s}(t) = 0.5 \tau_{\eta s}(t) \varepsilon_{\eta s}(t) \, \text{sgn}[\tau_{\eta s}(t), \varepsilon_{\eta s}(t)], \qquad (4.35)$$

$$W_\eta(t) = 0.5 \sigma_\eta(t) \varepsilon_\eta(t) \, \text{sgn}[\sigma_\eta(t), \varepsilon_\eta(t)] \qquad (4.36)$$

and sgn[x,y] is given by (4.13).

Finally, the following expression is obtained

$$W_{eq}(t) = \frac{\beta}{2}\tau_{\eta s}(t)\varepsilon_{\eta s}(t)\text{sgn}\left[\tau_{\eta s}(t),\varepsilon_{\eta s}(t)\right] + \frac{\kappa}{2}\sigma_{\eta}(t)\varepsilon_{\eta}(t)\text{sgn}\left[\sigma_{\eta}(t),\varepsilon_{\eta}(t)\right]. \quad (4.37)$$

Special cases of criterion (4.37) are discussed in the next chapters.

4.2.2 The Criterion of Maximum Parameter of Shear and Normal Strain Energy Density on the Critical Plane Determined by the Normal Strain Energy Density Parameter

The critical plane is defined by the normal strain energy density parameter. It is assumed that $Q = W_{af}$ and the critical plane with normal $\bar{\eta}$ is determined by normal loading, and position of the vector $\bar{s}$ is determined by one of the directions defined by the given shear loading.

From the strain and stress states for pure torsion, tension-compression or bending and under constant-amplitude loading, it is possible to derive relationships coupling the coefficients β and κ.

According to (4.34), the amplitude of the equivalent strain energy can be written as

$$W_{aeq} = \beta W_{a\eta s} + \kappa W_{a\eta}. \quad (4.38)$$

Under only normal loading the following expression is obtained

$$W_{aeq} = W_{a\eta} = \frac{\sigma_{axx}^2}{2E}(1 - \nu C), \quad (4.39)$$

where C is given by (2.3), and in the case of the sharp notch $C = \nu$.

For pure torsion on the plane of maximum tension, particular values of strain energy density are

$$W_{a\eta s} = 0 \quad (4.40)$$

and

$$W_{a\eta} = \frac{\sigma_{axx}^2}{2E}(1 - \nu C) \quad (4.41)$$

Introducing (4.39) – (4.41) to (4.38) subsequent expression is obtained

$$\kappa = 1. \qquad (4.42)$$

For bending on the plane of maximum tension, the same values like for torsion are obtained, so it is not possible to determine the coefficient β with an analytical method. This coefficient can be assorted depending on a material after non-proportional tests. It could be done for constant-amplitude fatigue tests with phase shift $\pi/2$. The final criterion form (4.34) takes the form

$$W_{eq}(t) = \beta W_{\eta s}(t) + W_{\eta}(t). \qquad (4.43)$$

4.2.3 The Criterion of Maximum Parameter of Shear and Normal Strain Energy Density in the Critical Plane Determined by the Shear Strain Energy Density Parameter

In this case, the critical plane is determined by the parameter of shear strain energy density. It is assumed that the critical plane with normal $\overline{\eta}$ and tangent $\overline{s}$ is defined as the mean position of one of two planes where the maximum shear strain energy density occurs. As in previous case, the equivalent parameter of strain energy density is determined from (4.34). Next, it is possible to write the amplitude of equivalent strain energy as a sum of parameters of normal and shear strain energy densities with weight coefficients β and κ.

Analysing the stress and strain state for pure torsion, tension-compression or bending under constant-amplitude loading, the relationships coupling the coefficients β and κ can be determined.

For pure torsion on the maximum shear plane, particular values of the strain energy density parameter are

$$W_{a\eta} = 0, \qquad (4.44)$$

$$W_{a\eta s} = 0.5\tau_{axy}\varepsilon_{axy} = 0.25\tau_{axy}\gamma_{axy} = 0.25\frac{\tau^2_{axy}}{G} = \frac{\tau^2_{axy}(1+\nu)}{2E} \qquad (4.45)$$

Introducing (4.41), (4.44), (4.45) into (4.38), the following formula is obtained

$$\frac{\sigma_{axx}^2(1-vC)}{2E} = \beta \frac{\tau_{axy}^2(1+v)}{2E}.$$ (4.46)

After transformation the following expression is obtained

$$\beta = \frac{\sigma_{axx}^2(1-vC)}{\tau_{axy}^2(1+v)}.$$ (4.47)

After introduction of

$$k = \frac{\sigma_{axx}^2}{\tau_{axy}^2},$$ (4.48)

the following expression can be obtained

$$\beta = \frac{k(1-vC)}{(1+v)}.$$ (4.49)

Similar calculations can be performed for pure bending or tension-compression on the maximum shear plane. Particular stresses and strains on the chosen plane are

$$\sigma_{axn} = \frac{\sigma_{axx} + \sigma_{ayy}}{2} = \frac{\sigma_{axx}(1+C)}{2},$$ (4.50)

$$\tau_{axy} = \frac{\sigma_{axx} - \sigma_{ayy}}{2} = \frac{\sigma_{axx}(1-C)}{2},$$ (4.51)

$$\gamma_{axy} = \frac{\varepsilon_{axx} - \varepsilon_{ayy}}{2} = \frac{\sigma_{axx} - v\sigma_{ayy} - \sigma_{ayy} + v\sigma_{axx}}{2E} = \frac{\sigma_{axx}(1-vC-C+v)}{2E},$$ (4.52)

$$\varepsilon_{axn} = \frac{\varepsilon_{axx} + \varepsilon_{ayy}}{2} = \frac{\sigma_{axx} - v\sigma_{ayy} + \sigma_{ayy} - v\sigma_{axx}}{2E} = \frac{\sigma_{axx}(1-vC+C-v)}{2E}.$$ (4.53)

4.2 The Energy Parameter Under Multiaxial Loading

Thus, the subsequent formula is obtained

$$W_{a\eta} = 0.5\sigma_{axn}\varepsilon_{axn} = 0.5\frac{\sigma_{axx}(1+C)}{2}\cdot\frac{\sigma_{axx}(1-vC+C-v)}{2E} = \frac{\sigma^2_{axx}(1-v)(1+C)^2}{8E}. \tag{4.54}$$

$$W_{a\eta s} = 0.5\tau_{axy}\gamma_{axy} = 0.5\frac{\sigma_{axx}(1-C)}{2}\cdot\frac{\sigma_{axx}(1-vC-C+v)}{2E} = \frac{\sigma^2_{axx}(1+v)(1-C)^2}{8E}. \tag{4.55}$$

Introducing (4.41), (4.54) and (4.55) into (4.38) the following expression is obtained

$$\frac{\sigma^2_{axx}}{2E}(1-vC) = \beta\frac{\sigma^2_{axx}(1+v)(1-C)^2}{8E} + \kappa\frac{\sigma^2_{axx}(1-v)(1+C)^2}{8E}. \tag{4.56}$$

After suitable transformations, the relation between the coefficients β and κ is obtained

$$(1-vC) = \beta\frac{(1+v)(1-C)^2}{4} + \kappa\frac{(1-v)(1+C)^2}{4}. \tag{4.57}$$

Thus

$$\kappa = \frac{4(1-vC)-\beta(1-C)^2(1+v)}{(1+C)^2(1-v)}. \tag{4.58}$$

After introduction of the determined value of the coefficient β the expression takes the form

$$\kappa = \frac{(4-k(1-C)^2)(1-vC)}{(1-v)(1+C)^2}. \tag{4.59}$$

If values of the coefficients β and κ are taken into account, (4.37) takes the form

$$W_{eq}(t) = \frac{k(1-\nu C)}{(1+\nu)} W_{\eta s}(t) + \frac{(4-k(1-C)^2)(1-\nu C)}{(1-\nu)(1+C)^2} W_{\eta}(t). \qquad (4.60)$$

If C = 0 for smooth elements, (4.60) for such elements takes the following form [155, 156, 187, 188]

$$W_{eq}(t) = \beta W_{\eta s}(t) + \frac{4-\beta(1+\nu)}{1-\nu} W_{\eta}(t). \qquad (4.61)$$

Relation between the coefficients β and κ and the ratio of normal stress amplitude (bending or tension-compression) to shear stress (torsion) for ν = 0.3 is shown in Fig. 4.3.

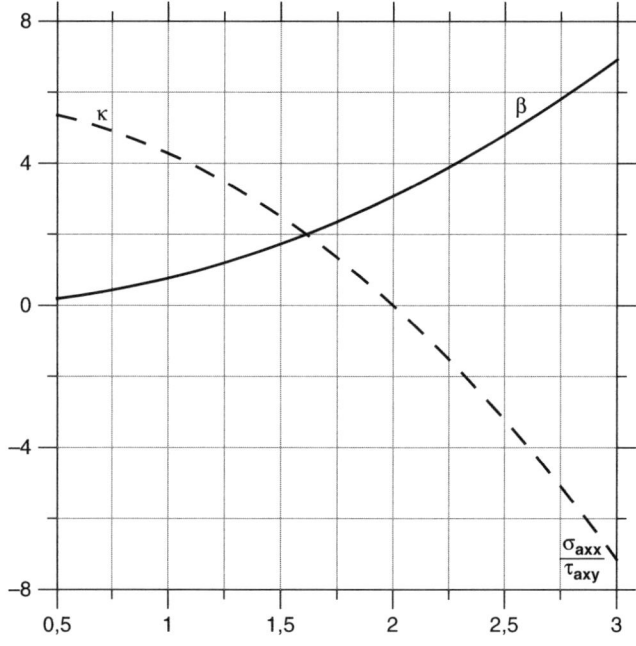

Fig. 4.3. Coefficients β and κ for different amplitude ratios σ_{axx}/τ_{axy}

4.3 Algorithm for Fatigue Life Assessment

Under multiaxial loading, fatigue life is calculated according to the general algorithm shown in Fig. 4.4. Particular stages of this algorithm are discussed below. Special attention was paid to the stages which were not previously presented in the algorithm describing procedures for uniaxial tension-compression in the stress description.

The proposed algorithms of fatigue life assessment for elements of machines and structures under multiaxial random loading have not been well verified so far and many laboratory tests should be performed in order to determine the ranges of their applicability in design calculations. Some of these algorithms have been already verified for chosen materials [132] and it is necessary to consider if they can be applied to calculation of fatigue life for other materials. Suitable selection of the multiaxial fatigue criteria seems be a very important problem. From a review of literature it appears that many proposed criteria of multiaxial fatigue are based on the critical plane. How should the critical plane be defined? The model presented below is based on the energy criteria that were presented earlier in this work.

Figure 4.5 shows the general algorithm of fatigue life determination using the criterion of shear and normal strain energy density parameter on the critical plane [115, 123, 136, 164].

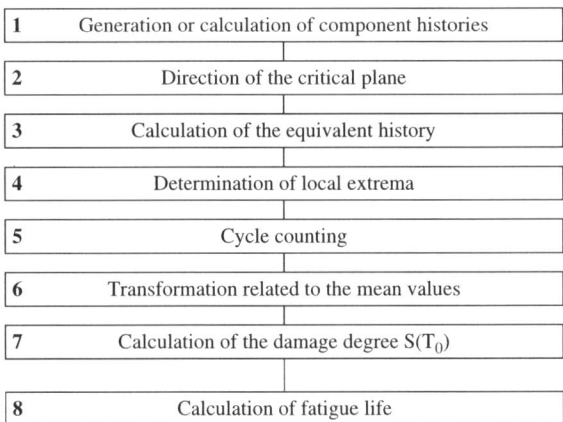

Fig. 4.4. Algorithm for fatigue life calculations under multiaxial random loading when the expected critical plane position is determined with the damage accumulation method

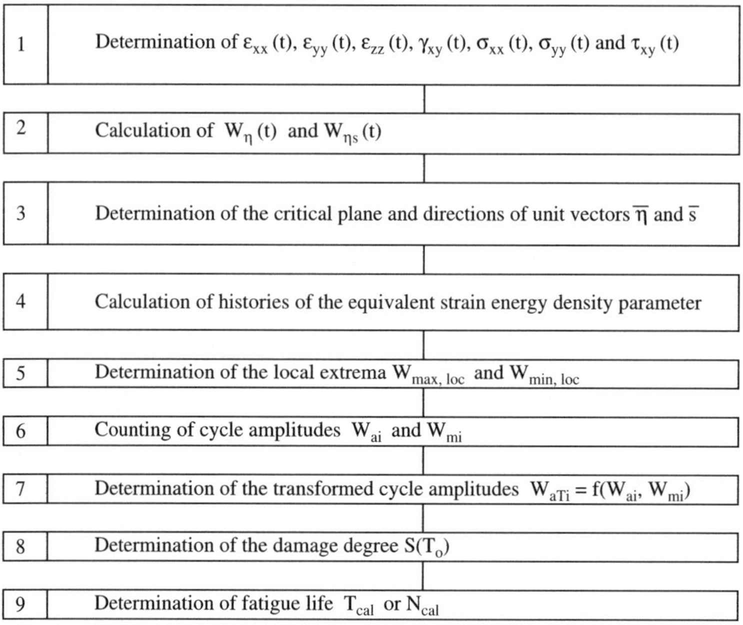

4 The Energy Model of Fatigue Life Assessment

1	Determination of $\varepsilon_{xx}(t), \varepsilon_{yy}(t), \varepsilon_{zz}(t), \gamma_{xy}(t), \sigma_{xx}(t), \sigma_{yy}(t)$ and $\tau_{xy}(t)$
2	Calculation of $W_\eta(t)$ and $W_{\eta s}(t)$
3	Determination of the critical plane and directions of unit vectors $\overline{\eta}$ and $\overline{s}$
4	Calculation of histories of the equivalent strain energy density parameter
5	Determination of the local extrema $W_{max,\,loc}$ and $W_{min,\,loc}$
6	Counting of cycle amplitudes W_{ai} and W_{mi}
7	Determination of the transformed cycle amplitudes $W_{aTi} = f(W_{ai}, W_{mi})$
8	Determination of the damage degree $S(T_o)$
9	Determination of fatigue life T_{cal} or N_{cal}

Fig. 4.5. Stages of fatigue life calculations under multiaxial loading according to the energy model

Stage 1

Histories of strain, $\varepsilon(t)$ and / or stress, $\sigma(t)$ are the input data for fatigue life calculations. As it was said in Sect. 3.1, they can come from measurements of actual strains or forces and moments, from previous numerical calculations or computer generation of random sequences with the formed probabilistic characteristics corresponding to service conditions or expected states. Such generations and simulation calculations are especially important under complex loading, when experimental tests are very expensive [92, 93, 126, 128, 237].

General models are based on the full matrix of stresses

$$\sigma_{ij} = \begin{bmatrix} \sigma_{xx} & \tau_{xy} & \tau_{xz} \\ \tau_{xy} & \sigma_{yy} & \tau_{yz} \\ \tau_{xz} & \tau_{yz} & \sigma_{zz} \end{bmatrix}. \tag{4.62}$$

4.3 Algorithm for Fatigue Life Assessment

In the considered case, the plane stress state occurs in the notch root

$$\sigma_{ij} = \begin{bmatrix} \sigma_{xx} & \tau_{xy} & 0 \\ \tau_{xy} & \sigma_{yy} & 0 \\ 0 & 0 & 0 \end{bmatrix}. \qquad (4.63)$$

It is important to note that in the critical place of stress concentration (in the considered edge of the weld) the local biaxial stress state with components σ_{xx}, σ_{yy} and τ_{xy} can be observed. These local components are obtained by nominal values $\sigma_{xx,n}(t)$ and $\tau_{xy,n}(t)$, and particular theoretical stress concentration factors by action of the fatigue notch

$$\sigma_{xx}(t) = K_{tb} \sigma_{xx,n}(t), \qquad (4.64)$$

$$\sigma_{yy}(t) = C \sigma_{xx}(t) \qquad (4.65)$$

and

$$\tau_{xy}(t) = K_{tt} \tau_{xy,n}(t). \qquad (4.66)$$

These stresses correspond to nominal strains which can be determined from the following relationships under elastic conditions:

$$\varepsilon_{xx,n}(t) = \frac{\sigma_{xx,n}(t)}{E}, \qquad (4.67)$$

$$\varepsilon_{yy,n}(t) = -\nu \frac{\sigma_{xx,n}(t)}{E}, \qquad (4.68)$$

$$\varepsilon_{zz,n}(t) = -\nu \frac{\sigma_{xx,n}(t)}{E}, \qquad (4.69)$$

$$\varepsilon_{xy,n}(t) = \frac{\gamma_{xy,n}(t)}{2} = \frac{\tau_{xy,n}(t)}{2G}. \qquad (4.70)$$

Generally, strains can be written as the full matrix of strains

$$\varepsilon_{ij} = \begin{bmatrix} \varepsilon_{xx} & \varepsilon_{xy} & \varepsilon_{xz} \\ \varepsilon_{xy} & \varepsilon_{yy} & \varepsilon_{yz} \\ \varepsilon_{xz} & \varepsilon_{yz} & \varepsilon_{zz} \end{bmatrix}. \qquad (4.71)$$

In the considered case, the plane stress state in the notch root can be observed, so for strains

$$\varepsilon_{ij} = \begin{bmatrix} \varepsilon_{xx} & \varepsilon_{xy} & 0 \\ \varepsilon_{xy} & \varepsilon_{yy} & 0 \\ 0 & 0 & \varepsilon_{zz} \end{bmatrix}. \qquad (4.72)$$

For local strains

$$\varepsilon_{xx}(t) = (1-Cv)K_{tb}\frac{\sigma_{xx,n}(t)}{E}, \qquad (4.73)$$

$$\varepsilon_{yy}(t) = (C-v)K_{tb}\frac{\sigma_{xx,n}(t)}{E}, \qquad (4.74)$$

$$\varepsilon_{zz}(t) = -(C+v)K_{tb}\frac{\sigma_{xx,n}(t)}{E}, \qquad (4.75)$$

$$\varepsilon_{xy}(t) = \frac{\gamma_{xy}(t)}{2} = \frac{K_{tt}\gamma_{xy,n}(t)}{2} = K_{tt}\frac{\tau_{xy,n}(t)}{2G}. \qquad (4.76)$$

For the sharp notch $C = v$ the $\varepsilon_{yy} = 0$ is obtained – it results from (4.74) and means that the plane state is accompanied with the plane strain state. On the surface on the notch edges (or weld edges) the stresses $\sigma_\eta(t)$ and $\tau_{\eta s}(t)$ can be calculated according to the following equations:

$$\sigma_\eta(t) = \hat{l}_\eta^{\,2}\sigma_{xx}(t) + \hat{m}_\eta^{\,2}\sigma_{yy}(t) + \hat{n}_\eta^{\,2}\sigma_{zz}(t) \\ + 2\hat{l}_\eta\hat{m}_\eta\sigma_{xy}(t) + 2\hat{l}_\eta\hat{n}_\eta\sigma_{xz}(t) + 2\hat{m}_\eta\hat{n}_\eta\sigma_{yz}(t) \qquad (4.77)$$

and

$$\tau_{\eta s}(t) = \hat{l}_\eta \hat{l}_s \sigma_{xx}(t) + \hat{m}_\eta \hat{m}_s \sigma_{yy}(t) + \hat{n}_\eta \hat{n}_s \sigma_{zz}(t) + (\hat{l}_\eta \hat{m}_s + \hat{l}_s \hat{m}_\eta)\sigma_{xy}(t) \\ + (\hat{l}_\eta \hat{n}_s + \hat{l}_s \hat{n}_\eta)\sigma_{xz}(t) + (\hat{m}_\eta \hat{n}_s + \hat{m}_s \hat{n}_\eta)\sigma_{yz}(t). \\ \qquad (4.78)$$

The strains $\varepsilon_\eta(t)$ and $\varepsilon_{\eta s}(t)$ are calculated from:

$$\varepsilon_\eta(t) = \hat{l}_\eta^2 \varepsilon_{xx}(t) + \hat{m}_\eta^2 \varepsilon_{yy}(t) + \hat{n}_\eta^2 \varepsilon_{zz}(t)$$
$$+ 2\hat{l}_\eta \hat{m}_\eta \varepsilon_{xy}(t) + 2\hat{l}_\eta \hat{n}_\eta \varepsilon_{xz}(t) + 2\hat{m}_\eta \hat{n}_\eta \varepsilon_{yz}(t) \quad (4.79)$$

and

$$\varepsilon_{\eta s}(t) = \hat{l}_\eta \hat{l}_s \varepsilon_{xx}(t) + \hat{m}_\eta \hat{m}_s \varepsilon_{yy}(t) + \hat{n}_\eta \hat{n}_s \varepsilon_{zz}(t) + (\hat{l}_\eta \hat{m}_s + \hat{l}_s \hat{m}_\eta)\varepsilon_{xy}(t) +$$
$$(\hat{l}_\eta \hat{n}_s + \hat{l}_s \hat{n}_\eta)\varepsilon_{xz}(t) + (\hat{m}_\eta \hat{n}_s + \hat{m}_s \hat{n}_\eta)\varepsilon_{yz}(t). \quad (4.80)$$

Because of the biaxial stress state on the surface of edges of the notch (or the weld), the stresses $\sigma_\eta(t)$ and $\tau_{\eta s}(t)$, and the strains $\varepsilon_\eta(t)$ and $\varepsilon_{\eta s}(t)$ –according to (4.77)–(4.78) – can be calculated from:

$$\sigma_\eta(t) = \hat{l}_\eta^2 \sigma_{xx}(t) + \hat{m}_\eta^2 \sigma_{yy}(t) + 2\hat{l}_\eta \hat{m}_\eta \sigma_{xy}(t), \quad (4.81)$$

$$\tau_{\eta s}(t) = \hat{l}_\eta \hat{l}_s \sigma_{xx}(t) + \hat{m}_\eta \hat{m}_s \sigma_{yy}(t) + (\hat{l}_\eta \hat{m}_s + \hat{l}_s \hat{m}_\eta)\sigma_{xy}(t), \quad (4.82)$$

$$\varepsilon_\eta(t) = \hat{l}_\eta^2 \varepsilon_{xx}(t) + \hat{m}_\eta^2 \varepsilon_{yy}(t) + \hat{n}_\eta^2 \varepsilon_{zz}(t) + 2\hat{l}_\eta \hat{m}_\eta \varepsilon_{xy}(t), \quad (4.83)$$

$$\varepsilon_{\eta s}(t) = \hat{l}_\eta \hat{l}_s \varepsilon_{xx}(t) + \hat{m}_\eta \hat{m}_s \varepsilon_{yy}(t) + \hat{n}_\eta \hat{n}_s \varepsilon_{zz}(t) + (\hat{l}_\eta \hat{m}_s + \hat{l}_s \hat{m}_\eta)\varepsilon_{xy}(t). \quad (4.84)$$

Stage 2
If the histories of stresses and strains in a plane defined by direction cosines are known, it is possible to define histories of the strain energy density parameter, using (4.81)–(4.84). It should be noted that in the further part of strain in direction of axis y, ε_{yy} are neglected as they are not significant because stresses in this direction are zero, $\sigma_{yy}=0$.

All the mentioned energy criteria of multiaxial fatigue – (4.43) and (4.61) – use the normal, W_η, and shear, $W_{\eta s}$, strain energy density parameter.

The normal strain energy density parameter according to (4.36), (4.81) and (4.83) can be written as:

$$W_\eta(t) = 0.5 \left[\hat{l}_\eta{}^2 \sigma_{xx}(t) + \hat{m}_\eta{}^2 \sigma_{yy}(t) + 2\hat{l}_\eta \hat{m}_\eta \tau_{xy}(t) \right] \cdot$$
$$\left[\hat{l}_\eta{}^2 \varepsilon_{xx}(t) + \hat{m}_\eta{}^2 \varepsilon_{yy}(t) + 2\hat{l}_\eta \hat{m}_\eta \varepsilon_{xy}(t) \right]$$
$$\operatorname{sgn}\{ \left[\hat{l}_\eta{}^2 \sigma_{xx}(t) + \hat{m}_\eta{}^2 \sigma_{yy}(t) + 2\hat{l}_\eta \hat{m}_\eta \tau_{xy}(t) \right],$$
$$\left[\hat{l}_\eta{}^2 \varepsilon_{xx}(t) + \hat{m}_\eta{}^2 \varepsilon_{yy}(t) + 2\hat{l}_\eta \hat{m}_\eta \varepsilon_{xy}(t) \right] \}, \qquad (4.85)$$

where stresses $\sigma_{xx}(t)$, $\sigma_{yy}(t)$, $\tau_{xy}(t)$ and strains $\varepsilon_{xx}(t)$, $\varepsilon_{yy}(t)$, $\gamma_{xy}(t)$ are the local values according to (4.66), (4.73), (4.74) and (4.76), respectively.

The shear strain energy density parameter – according to (4.35), (4.82) and (4.84) – takes the form:

$$W_{\eta s}(t) = 0.5 \left[\hat{l}_\eta \bar{l}_s \sigma_{xx}(t) + \hat{m}_\eta \hat{m}_s \sigma_{yy}(t) + \left(\hat{l}_\eta \hat{m}_s + \hat{l}_s \hat{m}_\eta \right) \tau_{xy}(t) \right] \cdot$$
$$\left[\hat{l}_\eta \bar{l}_s \varepsilon_{xx}(t) + \hat{m}_\eta \hat{m}_s \varepsilon_{yy}(t) + \left(\hat{l}_\eta \hat{m}_s + \hat{l}_s \hat{m}_\eta \right) \varepsilon_{xy}(t) \right] \cdot$$
$$\operatorname{sgn}\{ \left[\hat{l}_\eta \bar{l}_s \sigma_{xx}(t) + \hat{m}_\eta \hat{m}_s \sigma_{yy}(t) + \left(\hat{l}_\eta \hat{m}_s + \hat{l}_s \hat{m}_\eta \right) \tau_{xy}(t) \right],$$
$$\left[\hat{l}_\eta \bar{l}_s \varepsilon_{xx}(t) + \hat{m}_\eta \hat{m}_s \varepsilon_{yy}(t) + \left(\hat{l}_\eta \hat{m}_s + \hat{l}_s \hat{m}_\eta \right) \varepsilon_{xy}(t) \right] \}, \qquad (4.86)$$

where the stresses $\sigma_{xx}(t)$, $\sigma_{yy}(t)$, $\tau_{xy}(t)$ and the strains $\varepsilon_{xx}(t)$, $\varepsilon_{yy}(t)$, $\gamma_{xy}(t)$ are local values, according to (4.66), (4.73), (4.74) and (4.76).

Stage 3

In the algorithm for fatigue life assessment, proper determination of the expected position of the critical plane in the point of the maximum material effort is very important. The stress and strain states in the material belong to the basic factors determining this plane position. Its position is defined by the direction cosines $\hat{l}_n, \hat{m}_n, \hat{n}_n$ (n = η, s) of unit vectors $\bar{\eta}$ i $\bar{s}$ occurring in the fatigue criteria, where $\bar{\eta}$ is perpendicular, and $\bar{s}$ is tangent to the critical plane (Fig. 4.6), i.e.

$$\bar{\eta} \circ \bar{s} = 0. \qquad (4.87)$$

The following three methods are proposed for determination of the expected position of the critical position of fatigue fracture [109, 132, 161, 162]:

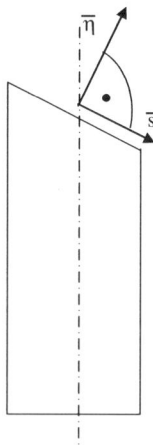

Fig. 4.6. Determination of the versors $\bar{\eta}$ and $\bar{s}$ defining the critical plane

A – *the weight function method*, presented in [163]. In this method, instantaneous values of angles $\alpha_n(t), \beta_n(t), \gamma_n(t)$ are averaged. These angles determine instantaneous positions of principal axes of strains or stresses with special weight functions. Modifications of this method apply Euler angle averaging and they can be found in [56, 61, 62, 69, 70, 71, 72].

B – *the maximum variance method*, applied in [109, 132]. In the variance method it is assumed that the planes where variance of the equivalent history according to the chosen fatigue criterion reaches its maximum, are critical for the material. This method does not require much time for calculations, but statistic parameters of the strain (or stress) tensor components should be known. The variance method can be effective if the stress tensor components are stationary and ergodic stochastic processes with the same statistic character of loading.

C – *the method of damage accumulation*, discussed in [63, 67, 68, 101, 120, 133, 142]. Here, fatigue damages are accumulated in many planes of the given material point and the plane where the damage degree is maximum (i.e. where fatigue life is minimum) is selected. In the case when the assumed direction of the critical plane coincides with the criterion applied for fatigue life calculation, not only the expected critical plane direction is obtained but life as well. This method is more and more applied at present.

In the case of the method of damage accumulation and the variance method, their success depends on selection of a proper fatigue criterion and a step of angle change discretization. The optimisation methods accelerating determination of the expected position of the critical plane are also applied. The critical plane and the fracture plane should be, however, distinguished. They are not always the same planes. Figure 4.7 shows the cracking models. In model I, normal stresses are responsible for fatigue cracking, in models II and III shear stresses cause cracking. There are materials where both shear and normal stresses are responsible for cracking. Then, it can be defined as the mixed model. The elastic-brittle materials crack according to model I, and the elastic-plastic materials – according to models II or III.

The model of fatigue damage accumulation used in this paper was verified many times for stress, strain and energy models.

Direction cosines $\hat{l}_\eta, \hat{m}_\eta, \hat{l}_s, \hat{m}_s$ of vectors $\hat{\eta}$ and $\hat{s}$, occurring in formulas for the normal or shear strain energy density parameter, (4.85) and (4.86), are under the plane stress state defined by one angle α in the following relationships:

$$\hat{l}_\eta = \cos\alpha, \quad \hat{m}_\eta = \sin\alpha, \quad \hat{l}_s = -\sin\alpha, \quad \hat{m}_s = \cos\alpha. \quad (4.88)$$

Under random or variable-amplitude loading, a damage degree is calculated according to the algorithm shown in Fig. 4.8, but it is necessary to assume a suitable criterion of normal or shear strain energy density parameter, (4.85) or (4.86). From the previous considerations [156] it appears that the criterion assuming the plane determined by the maximum parameter of normal strain energy density, as the critical plane is valid for cast iron, i.e. a cast brittle material. Whereas, the criterion defined on the plane of the maximum parameter of shear strain energy density should be applied in life calculations for steels and non-ferrous metal alloys.

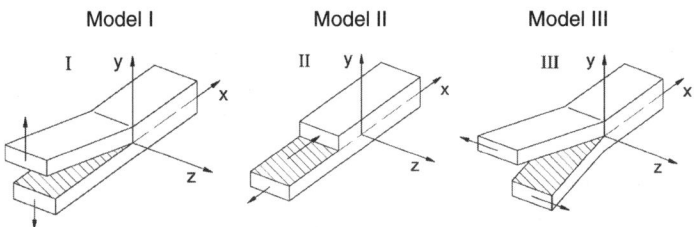

Fig. 4.7. Models of fatigue cracking [23]

4.3 Algorithm for Fatigue Life Assessment 55

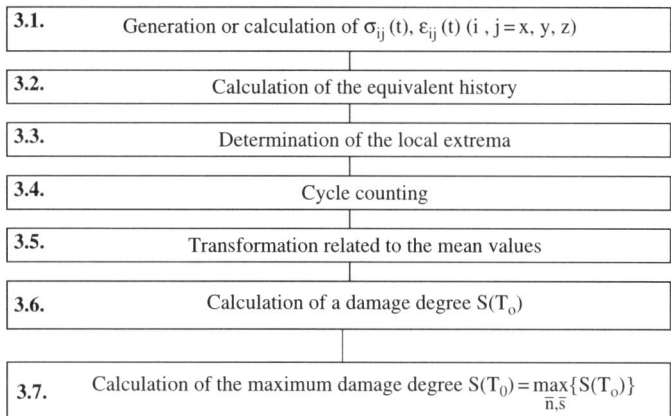

Fig. 4.8. Algorithm for determination of the critical plane direction by the normal or shear strain energy density parameter with the method of damage accumulation

The above algorithm becomes simplified under constant-amplitude loading. In such a case, it is necessary to determine the energy parameter history only for one cycle (one period T). If the normal strain energy density parameter is applied for determination of the critical plane, $W_\eta(t)$ for one cycle only should be calculated. Then, amplitude of the normal strain energy density parameter $W_{\eta a}$ is

$$W_{eq,a} = W_{\eta a} = \max_{T,\alpha} W_\eta(t,\alpha). \tag{4.89}$$

Equation (4.89) defines a position of the critical fracture plane expressed by $\alpha_0 = \alpha$. The critical plane is defined by the maximum equivalent amplitude of the normal strain energy density parameter $W_{eq,a}$.

If the shear strain energy density parameter is applied for determination of the critical plane, it is necessary to calculate $W_{\eta s}(t)$ for one cycle. Then, amplitude of the shear strain energy density parameter is $W_{\eta sa}$, and the shear strain energy density parameter is

$$W_{eq,a} = W_{\eta sa} = \max_{T,\alpha} W_{ns}(t,\alpha). \tag{4.90}$$

As in the case of the normal strain energy density parameter, Eq. (4.90) defines a position of the critical fracture plane expressed as $\alpha_0 = \alpha$. The critical plane is defined by the maximum equivalent amplitude of the shear strain energy density parameter $W_{eq,a}$.

Stage 4
When the critical plane position is determined, it is necessary to determine the equivalent parameter of strain energy density on the critical plane.

In the case of the critical plane defined by the shear strain energy density parameter (4.86), the equivalent strain energy density parameter (4.61) can be defined as

$$W_{eq}(t) = 0.5 \left[\hat{l}_\eta^2 \sigma_{xx}(t) + \hat{m}_\eta^2 \sigma_{yy}(t) + 2\hat{l}_\eta \hat{m}_\eta \tau_{xy}(t)\right] \cdot$$
$$\left[\hat{l}_\eta^2 \varepsilon_{xx}(t) + \hat{m}_\eta^2 \varepsilon_{yy}(t) + 2\hat{l}_\eta \hat{m}_\eta \varepsilon_{xy}(t)\right]$$
$$\text{sgn}\left\{\left[\hat{l}_\eta^2 \sigma_{xx}(t) + \hat{m}_\eta^2 \sigma_{yy}(t) + 2\hat{l}_\eta \hat{m}_\eta \tau_{xy}(t)\right],\right.$$
$$\left.\left[\hat{l}_\eta^2 \varepsilon_{xx}(t) + \hat{m}_\eta^2 \varepsilon_{yy}(t) + 2\hat{l}_\eta \hat{m}_\eta \varepsilon_{xy}(t)\right]\right\}$$
$$+ 0.5\beta \left[\hat{l}_\eta \bar{l}_s \sigma_{xx}(t) + \hat{m}_\eta \hat{m}_s \sigma_{yy}(t) + \left(\hat{l}_\eta \hat{m}_s + \hat{l}_s \hat{m}_\eta\right) \tau_{xy}(t)\right] \cdot$$
$$\left[\hat{l}_\eta \bar{l}_s \varepsilon_{xx}(t) + \hat{m}_\eta \hat{m}_s \varepsilon_{yy}(t) + \left(\hat{l}_\eta \hat{m}_s + \hat{l}_s \hat{m}_\eta\right) \varepsilon_{xy}(t)\right] \cdot$$
$$\text{sgn}\left\{\left[\hat{l}_\eta \bar{l}_s \sigma_{xx}(t) + \hat{m}_\eta \hat{m}_s \sigma_{yy}(t) + \left(\hat{l}_\eta \hat{m}_s + \hat{l}_s \hat{m}_\eta\right) \tau_{xy}(t)\right],\right.$$
$$\left.\left[\hat{l}_\eta \bar{l}_s \varepsilon_{xx}(t) + \hat{m}_\eta \hat{m}_s \varepsilon_{yy}(t) + \left(\hat{l}_\eta \hat{m}_s + \hat{l}_s \hat{m}_\eta\right) \varepsilon_{xy}(t)\right]\right\}. \quad (4.91)$$

In the case of the critical plane defined by the normal strain energy density parameter (4.85), the equivalent strain energy density parameter (4.43) can be expressed as

$$W_{eq}(t) = 0.5 \frac{4 - \beta(1+\nu)}{1-\nu} \left[\hat{l}_\eta^2 \sigma_{xx}(t) + \hat{m}_\eta^2 \sigma_{yy}(t) + 2\hat{l}_\eta \hat{m}_\eta \tau_{xy}(t)\right] \cdot$$
$$\left[\hat{l}_\eta^2 \varepsilon_{xx}(t) + \hat{m}_\eta^2 \varepsilon_{yy}(t) + 2\hat{l}_\eta \hat{m}_\eta \varepsilon_{xy}(t)\right]$$
$$\text{sgn}\left\{\left[\hat{l}_\eta^2 \sigma_{xx}(t) + \hat{m}_\eta^2 \sigma_{yy}(t) + 2\hat{l}_\eta \hat{m}_\eta \tau_{xy}(t)\right],\right.$$
$$\left.\left[\hat{l}_\eta^2 \varepsilon_{xx}(t) + \hat{m}_\eta^2 \varepsilon_{yy}(t) + 2\hat{l}_\eta \hat{m}_\eta \varepsilon_{xy}(t)\right]\right\}$$
$$+ 0.5\beta \left[\hat{l}_\eta \bar{l}_s \sigma_{xx}(t) + \hat{m}_\eta \hat{m}_s \sigma_{yy}(t) + \left(\hat{l}_\eta \hat{m}_s + \hat{l}_s \hat{m}_\eta\right) \tau_{xy}(t)\right] \cdot$$
$$\left[\hat{l}_\eta \bar{l}_s \varepsilon_{xx}(t) + \hat{m}_\eta \hat{m}_s \varepsilon_{yy}(t) + \left(\hat{l}_\eta \hat{m}_s + \hat{l}_s \hat{m}_\eta\right) \varepsilon_{xy}(t)\right] \cdot$$

$$\text{sgn}\{[\hat{l}_\eta \bar{l}_s \sigma_{xx}(t) + \hat{m}_\eta \hat{m}_s \sigma_{yy}(t) + (\hat{l}_\eta \hat{m}_s + \hat{l}_s \hat{m}_\eta)\tau_{xy}(t)],$$
$$[\hat{l}_\eta \bar{l}_s \varepsilon_{xx}(t) + \hat{m}_\eta \hat{m}_s \varepsilon_{yy}(t) + (\hat{l}_\eta \hat{m}_s + \hat{l}_s \hat{m}_\eta)\varepsilon_{xy}(t)]\}, \quad (4.92)$$

where β is defined by (4.49), and for sharp notches (C=v) the following expression is obtained

$$\beta = k(1+v) \quad (4.93)$$

and after introduction of (4.48) into (4.93) it takes the following form

$$\beta = \left(\frac{\sigma_{axx}(N_f)}{\tau_{axy}(N_f)}\right)^2 (1+v), \quad (4.94)$$

where $\sigma_{axx}(N_f)$ and $\tau_{axy}(N_f)$ are fatigue strength under simple loading states (tension-compression (bending) and shearing (torsion), respectively) versus a number of cycles. These two fatigue characteristics are often parallel. Then, it is suitable to substitute the fatigue limit, and finally β from (4.94) is obtained

$$\beta = \left(\frac{\sigma_{af}}{\tau_{af}}\right)^2 (1+v). \quad (4.95)$$

Stage 5
At this stage, extrema of the energy parameter history are determined from (4.91) and (4.92), like in the case of stress histories.

Stage 6
As it was said, the rain flow method (the envelope method) allows to determine both cycles and half-cycles. They are determined by suitable envelopes (see Fig. 4.2). This method is programmed and cycles are counted by a computer. Each time, the amplitude and the mean value of a cycle or a half-cycle are determined. Numerical procedure of cycle counting has been shown in Fig. 4.3. The same procedure is proposed for the strain energy density parameter.

Stage 7
At this stage, cycle amplitudes of the energy parameter are transformed in relation to the occurring mean values of this parameter. Such model was presented in [77, 78, 79]; however, it is not applied and thus not discussed in is paper.

Stage 8
There are many hypotheses concerning accumulation of amplitudes of stress cycles and half-cycles. As in the case of (3.21), it is possible to propose fatigue damage accumulation for cycle amplitudes of the strain energy density parameter [55, 58, 59, 64, 65, 66, 102, 103, 105, 106, 116, 122, 136, 154, 179, 180] according to a general formula

$$S(T_o) = \begin{cases} \sum_{i=1}^{j} \dfrac{n_i}{b * N * (W_{af}/W_{ai})^{m_w'}} & \text{for } W_{ai} \geq aW_{af} \\ h \sum_{i=j+1}^{k} \dfrac{n_i}{N_o (W_{af}/W_{ai})^{m_{1w}'}} & \text{for } W_{ai} < aW_{af} \end{cases}, \qquad (4.96)$$

where:
a – coefficient allowing to include amplitudes below W_{af} in damage accumulation,
m'_w – coefficient of the S–N curve slope for the strain energy density parameter,
W_{af} – fatigue limit according to the strain energy density parameter,
n_i– number of cycles with amplitude W_{ai} (two identical half-cycles form one cycle),
and the remaining notations – as in the stress model (3.20).

Stage 9
After determination of the damage degree at observation time T_o according to (4.96), fatigue life is calculated

$$T_{cal} = \frac{T_o}{S(T_o)} \qquad (4.97)$$

or

$$N_{cal} = \frac{N_{block}}{S(N_{block})} \qquad (4.98)$$

as for the uniaxial loading state.

5 An Example of Fatigue Life Evaluation Under Simple Loading

5.1 Fatigue Tests

Static properties of the considered materials are shown in Table 5.1. Their tests have been discussed in [224] and other papers. The tests were performed under uniaxial constant amplitudes, variable amplitudes of normal distribution (Gaussian spectrum) without and with overload under axial and bending loading with two stress ratios

$$R = \frac{\sigma_{min}}{\sigma_{max}}, \qquad (5.1)$$

under a symmetric cycle (R = –1) or pulsating loading (R = 0). Two kinds of welds were tested (Fig. 5.1). All the joints were made by one person. Sheets 1250 mm in length were joined with the GMAW method; 6–21 layers were put on, depending on the native material and its thickness. The specimens were cut with the plasma method, radii and transition angles in the notch root were determined for 12 specimens and 4 sections.

The theoretical notch coefficient for welded joints depends on the sheet thickness, the radius in the notch root and the angle of weld face inclination. Procedure of determination of theoretical notch coefficients is presented in many papers, for example [182, 183, 185].

Table 5.1. Static properties of steels

Steel	E, GPa	R_m, MPa	$R_{0.2}$, MPa
S355N	206	560	378
S355M	206	524	422
S680Q	206	868	784
S960Q	206	1072	998

5 An Example of Fatigue Life Evaluation Under Simple Loading

The theoretical notch coefficient for butt welds (Fig. 5.2) can be determined from

$$K_t = \left[1 + b_1 \left(\frac{t}{\rho}\right)^{b_2}\right]\left[1 + \left(a_o + a_1 \sin\theta + a_2 \sin^2\theta + a_3 \sin^3\theta\right)\left(\frac{t}{\rho}\right)^{l_1 + l_2 \sin(\theta + l_3)}\right],$$

(5.2)

The constants from (5.2) are presented in Table 5.2.

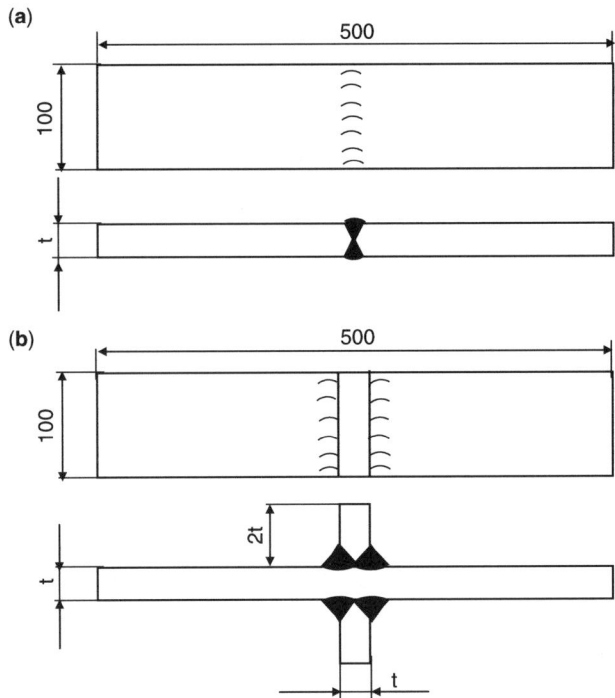

Fig. 5.1. Geometry of the tested welded joints: (**a**) butt welds, (**b**) fillet weld with the transverse stiffeners

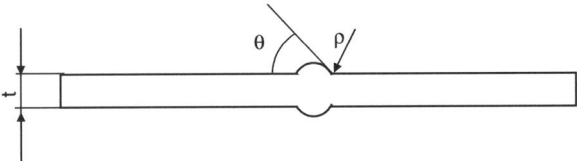

Fig. 5.2. Geometry of the butt welds joint for determination of the theoretical notch coefficient

Table 5.2. Coefficients in (5.2) depending on loading [8]

Coefficient Loading	a_o	a_1	a_2	a_3	b_1	b_2	l_1	l_2	l_3
axial	0.169	1.503	−1.968	0.713	−0.138	0.213	0.249	0.356	6.194
bending	0.181	1.207	−1.737	0.689	−0.156	0.207	0.292	0.349	3.283

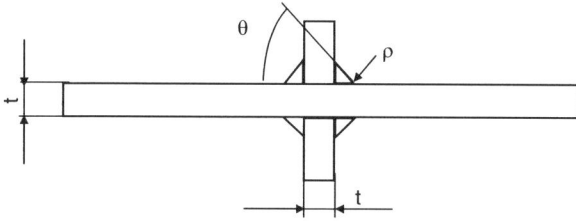

Fig. 5.3. Geometry of the fillet weld with the transverse stiffeners for determination of the theoretical notch coefficient

For the fillet joint with the transverse stiffeners (Fig. 5.3), the theoretical notch coefficient can be determined from the following equation

$$K_t = m_o + \left[1 + m_2\left(\frac{t}{\rho}\right)^{p_3} + m_3(\sin\theta)^{p_4}\right](\sin\theta)^{p_5}\left(\frac{t}{\rho}\right)^{p_6}. \qquad (5.3)$$

The constants from (5.3) are given in Table 5.3.

Next, coefficients of stress concentration K_t were determined and after statistical processing they were presented in Table 5.4. The specimens of 30 mm thickness were tested under bending, and the specimens of 10 mm thickness were tested under axial loading (tension-compression).

Table 5.3. Coefficients in (5.3) depending on loading

Coefficient Loading	m_o	m_2	m_3	p_3	p_4	p_5	p_6
axial	1.538	1.455	−2.933	0.208	1.213	2.086	0.207
bending	1.256	12.153	−3.738	0.154	0.481	1.723	0.172

Table 5.4. Coefficients of stress concentration K_t

Steel	S355N	S355M	S690Q	S960Q
transverse stiffeners (t = 10 mm)	4.78	3.67	3.79	4.27
butt welds (t = 10 mm)	2.06	2.47	2.69	2.22
transverse stiffeners (t = 30 mm)	5.20	6.03	4.24	6.20
butt welds (t = 30 mm)	2.82	3.13	2.67	3.02

5.1.1 Tests Under Constant-amplitude Loading

After analysis of experimental data obtained under uniaxial cyclic loading, all the data obtained under axial loading and plane bending were collected in two separate groups. From the following equations

$$\sigma_a = K_{ta}\sigma_{an} \qquad (5.4)$$

and

$$\sigma_a = K_{tb}\sigma_{an} \qquad (5.5)$$

the amplitudes of pseudoelastic local stress for axial loading and plane bending are obtained, respectively. Transition from the nominal system to the local system using the theoretical coefficient of stress concentration has been shown in Fig. 2.6.

The ASTM standard [9] was applied for all the materials and both kinds of welded joints. As a consequence, the following regression equations S–N in the Basquin notation (3.2) for axial loading is obtained

$$\lg N = 11.390 - 2.280 \lg \sigma_a \qquad \text{for} \quad R=-1 \qquad (5.6)$$

and

$$\lg N = 11.800 - 2.483 \lg \sigma_a \qquad \text{for} \quad R=0. \qquad (5.7)$$

Table 5.5 presents test scatters for particular kinds of loading for the significance level $\alpha = 5\%$ and for two standard deviations. From analysis of the data in Table 5.5 it appears that the mean scatter band for cyclic tests has the coefficient close to 4. Interpretation of the data is shown in Figs. 5.4 and 5.5. From the figures it appears that all the materials both kinds of welded joints for particular stress ratios (R = –1) and (R = 0) can be approximated with use of one fatigue curve S–N.

Table 5.5. Scaters of fatigue tests in relation to the S–N curves for simple loading states

No	Type of loading	R	sT_N	$T_N (\alpha=5\%)$	$T_N (2s)$
1	Axial loading	−1	1.919	3.878	3.838
2	Axial loading	0	2.133	4.266	4.266
3	Plane bending	−1	2.208	4.579	4.416
4	Plane bending	0	1.581	3.257	3.162
5	On the average		1.710	3.995	3.920

Here, sT_N is standard deviation of the scatters.

Similar analysis can be done for the plane bending

$$\lg N = 12.794 - 2.627 \lg \sigma_a \quad \text{for} \quad R=-1 \quad (5.8)$$

and

$$\lg N = 13.304 - 2.835 \lg \sigma_a \quad \text{for} \quad R=0. \quad (5.9)$$

Under axial loading from (5.6) and (5.7), the fatigue limit for pseudoelastic amplitudes of local stresses are σ_{afa} = 172 MPa and 176 MPa for R = −1 and R = 0 respectively for $N = 2 \cdot 10^6$ cycles. Figures 5.6 and 5.7 show the

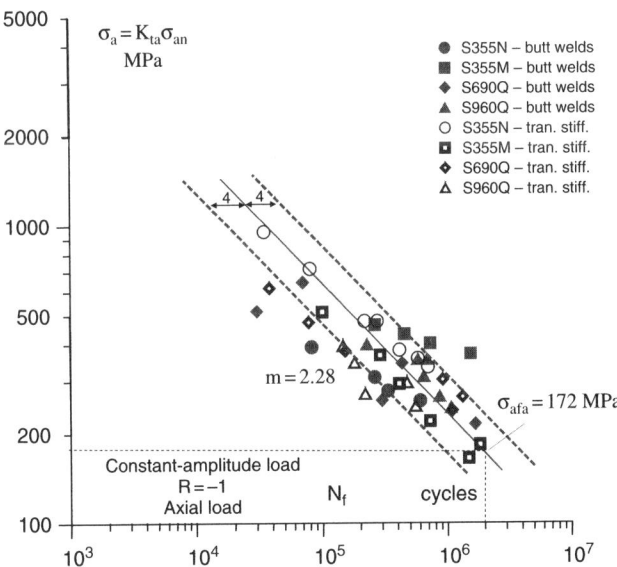

Fig. 5.4. Fatigue curves S–N for axial constant-amplitude loading for the considered materials and specimens under symmetric loading

test results for bending according to (5.8) (R = −1) and (5.9) (R = 0). Under plane bending from (5.7), The fatigue limit for pseudoelastic local stresses is $\sigma_{afb} = 296$ MPa under $N = 2 \cdot 10^6$ cycles. Fatigue strength does not depend on the material, but on a loading type (axial or bending). In the local stress system, influence of geometry cannot be seen, either. From Fig. 5.4 – (5.6), (5.7), Fig. 5.6 – (5.8) and Fig. 5.7 – (5.9) it results that the mean stress does not influence the fatigue life, probably because of the existing high residual stress, found during other measurements shown in [224].

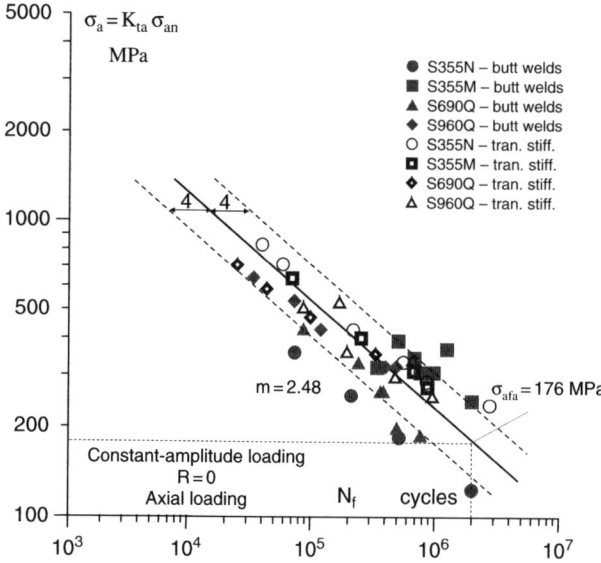

Fig. 5.5. Fatigue curves S–N for axial constant-amplitude loading for all the materials and specimens under pulsating loadings

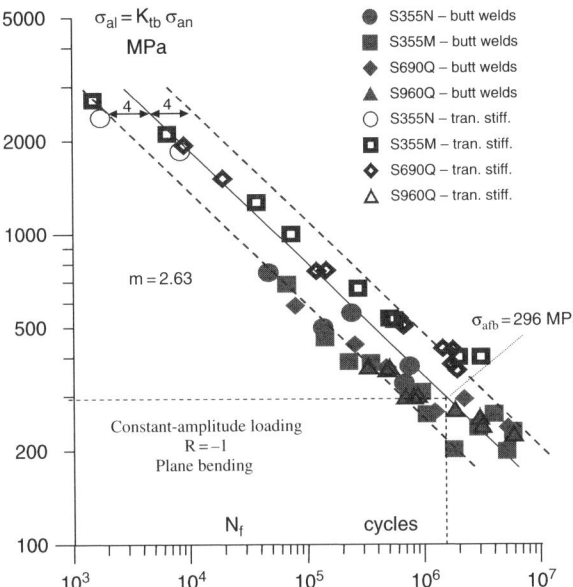

Fig. 5.6. Fatigue curves S–N for cyclic plane bending for the tested materials and specimens under symmetric loading

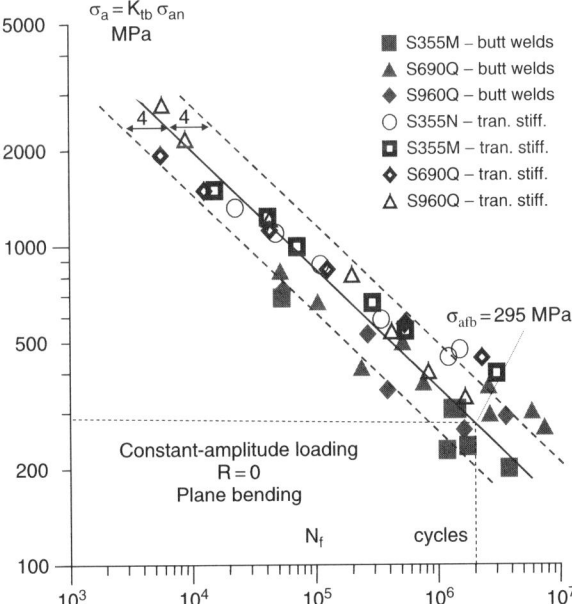

Fig. 5.7. Fatigue curves S–N for cyclic plane bending for the tested materials and specimens under pulsating loading

5.1.2 Tests Under Variable-amplitude Loading

In the case of Gaussian distribution of cyclic load amplitudes, the sequence length is $N_{block} = 5 \cdot 10^4$ cycles. In the case of the overload, overloads with a number of cycles $N_{OL} = 10^3$ are randomly distributed in the basic band, and the total spectrum length is $5 \cdot 10^4$ cycles, like for loading without overloads. As for the generated overloads, the ratio of their maximum values to the maximum value in the block of Gaussian loads is 1.4. Figure 5.8 shows a sequence of variable-amplitude loads with and without overloads versus the accumulated number of cycles n_{ic}. This sequence concerns normalized loads in Fig. 5.8a (symmetric loading) and Fig. 5.8b (pulsating loading).

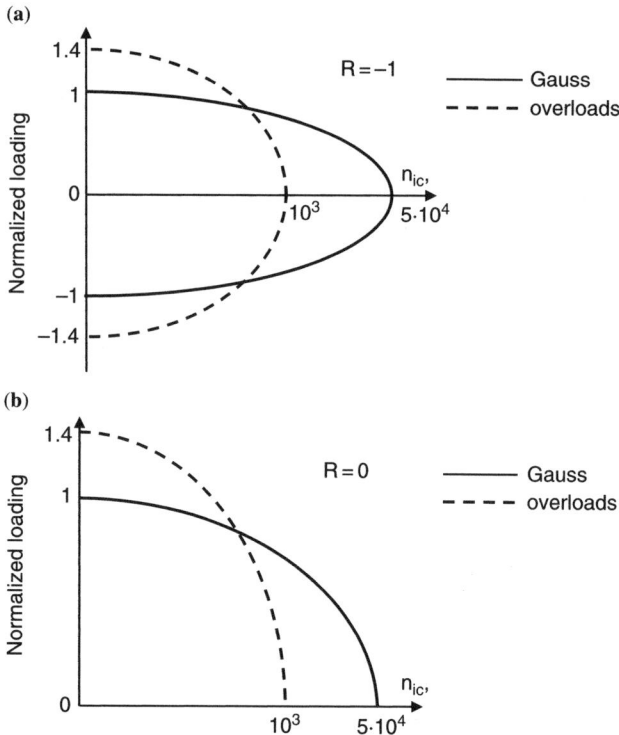

Fig. 5.8. Sequence of variable-amplitude loading with and without overloads, (**a**) symmetric loading, (**b**) pulsating loading versus the accumulated number of cycles n_{ic}

5.2 Verification of the Results Obtained Under Variable-amplitude Loading

Calculations were performed in order to compare calculation and experimental lives. The first such calculations were presented in [112, 228]. As it was said, large scatters of the fatigue test results were found. The scatters for cyclic tests were determined in the previous chapter. From the obtained data it appears that the scatter band of the cyclic test results has the coefficient 4. The calculated and experimental fatigue lives are presented in Figs. 5.9, 5.10 and 5.11. The calculation results are shown in Table 5.6. Many results are included into the scatter band with coefficient $T_N = 3$, but more of them are included into the scatter band with coefficient $T_N = 4$, but around the mean damage $\bar{T}_N = 1$. It means that the sum of most actual damages is included into the band

$$1/T_N < \bar{T}_N \approx 1 < T_N. \tag{5.10}$$

It concerns more than 95% of the considered data.

Table 5.6. Scatters of the fatigue test results related to S–N curves for simple loading states

| No | Type of loading | R | \multicolumn{2}{c}{Cyclic tests} | \multicolumn{2}{c}{Variable-amplitude tests} |
			$\bar{T}_N$	T_N (2s)	$\bar{T}_N$	T_N (2s)
1	Axial loading	−1	1	3.838	1.018	4.169
2	Plane bending	−1	1	4.416	1.143	4.385
3	Plane bending	0	1	3.162	1.449	3.750

68   5 An Example of Fatigue Life Evaluation Under Simple Loading

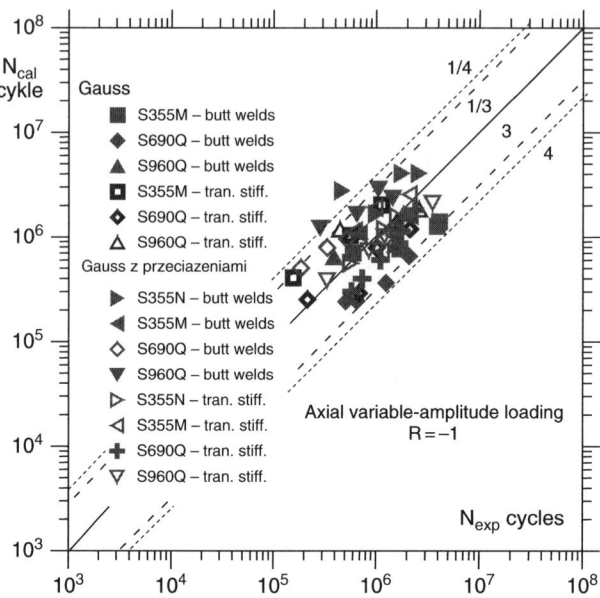

Fig. 5.9. Comparison of calculated and experimental fatigue lives for variable-amplitude axial loading

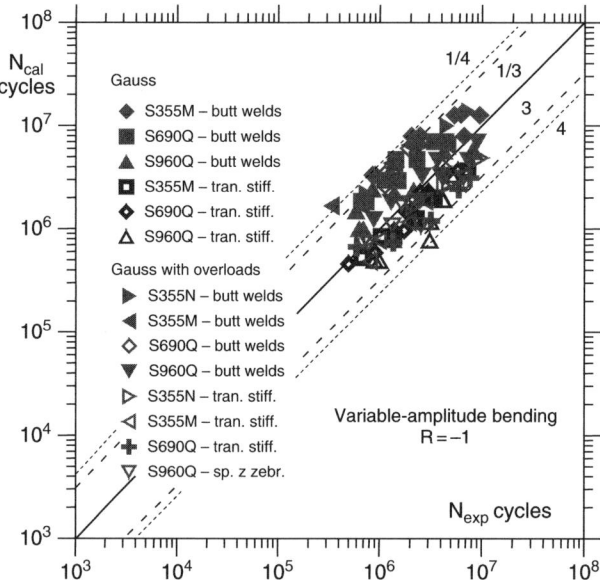

Fig. 5.10. Comparison of calculated and experimental fatigue lives for uniaxial variable-amplitude bending

5.2 Verification of the Results Obtained Under Variable-amplitude Loading 69

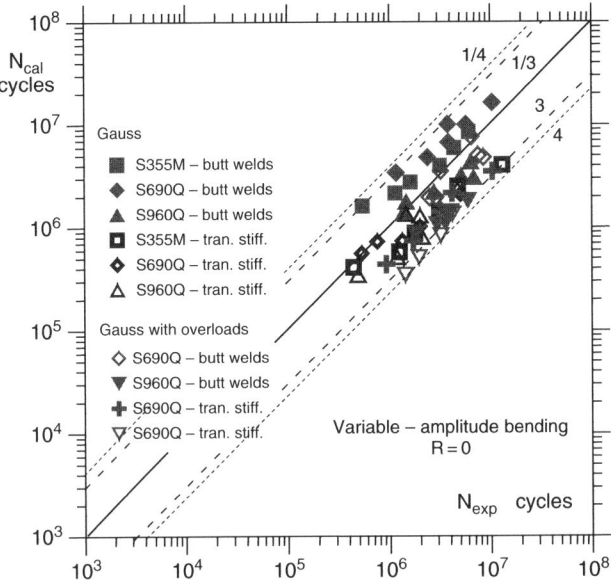

Fig. 5.11. Comparison of calculated and experimental fatigue lives for uniaxial variable-amplitude bending

6 An Example of Fatigue Life Evaluation Under Complex Loading States

6.1 Fatigue Tests

Experimental verification was based on fatigue tests performed by Sonsino [158, 165, 217, 220, 221, 227], Witt [240, 241, 242, 243] and Küppers [86, 87, 88, 113, 114, 226]. Table 6.1 presents mechanical properties of the considered materials (steel StE460 and aluminium alloy AlSi1MgMn T6). Figure 6.1 shows geometries of the tested specimens.

Welded joints were tested under pure bending, pure torsion and bending with torsion, in- and out-of-phase (90°). Under combined bending with torsion, a ratio of nominal shearing to bending stress was determined for tests performed by Sonsino and Küppers

$$\frac{\tau_{an}}{\sigma_{an}} = 0.58 \tag{6.1}$$

and for tests by Witt

$$\frac{\tau_a}{\sigma_a} = \frac{K_{tt}\tau_{an}}{K_{tb}\sigma_{an}} = 0.6. \tag{6.2}$$

Table 6.1. Mechanical properties of the tested materials

Material	E GPa	ν	$R_{0.2}$ MPa	R_m MPa	A_5 %
StE460 (Sonsino)	206	0.30	520	670	25
StE460 (Witt)	192	0.30	466	624	28.7
AlSi1MgMn T6 (6082 T6) (Küppers)	71.3	0.32	315	332	13

Table 6.2. The calculated radii and stress concentration factors for notches

Type of welded joint	ρ_{real}, mm	K_{tb}	K_{tt}
Flange-tube a (Sonsino)	0.45	3.93	1.85
Tube-tube (Sonsino)	0.45	2.42	1.77
Flange- tube (Witt)	no data	2.20	1.32
Flange-tube (Küppers)	17	1.62	1.14
Tube-tube (Küppers)	1.7	1.68	1.21

where ρ_{real} is an actual mean radius in the notch root.

Analysis of tests performed by Sonsino was done for rough specimens, and in the case of Witt's tests machined specimens were considered. Thus, the notch coefficients for Witt's tests are lower than those for Sonsino's tests (see Table 6.2). Table 6.2 shows comparison of all the notch coefficients for fictitious radii and stress concentration factors for actual radii. All the coefficients were defined by calculations with the finite element method for particular radii and constant angle of the weld face [217].

The stress concentration factors included in Table 6.2 confirm the known relationship [219, 225, 249]

$$K_{ta} > K_{tb} > K_{tt}. \tag{6.3}$$

In the case of rough welded joints, the stress concentration factor is a function of the radius in the notch root. From calculations, the relationships for tube-flange joints for bending were obtained

$$\log K'_{tb} = 0.505 - 0.267 \log \rho \tag{6.4}$$

and torsion

$$\log K'_{tt} = 0.215 - 0.151 \log \rho. \tag{6.5}$$

Similar relationships were obtained for tube-tube joints for bending and torsion, respectively

$$\log K'_{tb} = 0.299 - 0.235 \log \rho \tag{6.6}$$

Table 6.3. The calculated fictitious radii and stress concentration factors for notches

Type of joint	ρ^*, mm	ρ_{fb}, mm	K'_{fb}	ρ_{ft}, mm	K'_{ft}
Flange-tube	4.0	1.16	3.11	0.40	1.88
	3.5	1.02	3.21	0.35	1.92
Tube-tube	4.0	1.16	1.92	0.40	1.79
	3.5	1.02	1.98	0.35	1.83

and

$$\log K'_{tt} = 0.181 - 0.181 \log \rho. \tag{6.7}$$

Then, stress concentration factors were determined for bending and torsion, the substitute value of microstructure $\rho^* = 0.4$ mm was assumed. Such assumption is often made for constructional steels. Additional calculations were performed for $\rho^* = 0.35$ mm, according to Fig. 2.9. Next, according to (2.12) and the Huber-Mises-Hencky hypothesis, fictitious radii ρ_f were determined (see Table 6.3). From the data presented in Table 6.3 it appears that influence of $\rho^* = 0.35$ mm and 0.4 mm on determination of stress concentration factors is relatively small. Thus, the universal value $\rho^* = 0.4$ mm is recommended. From the same table it also results that different fictitious notch radii are obtained for bending and torsion.

After the recalculation of the values from the nominal system to the local system with the use of theoretical notch coefficients for bending and torsion a new S–N characteristics was obtained, i.e.

$$\log N_f = A - m \cdot \log \sigma_a \tag{6.8}$$

and

$$\log N_f = A_\tau - m_\tau \cdot \log \tau_a, \tag{6.9}$$

determined on the basis of the test results through regression analysis according to the ASTM standard [9]. The determined parameters of the fatigue curves S–N are given in Table 6.4, where m is inclination, and k is determined as

$$k(N_f) = \frac{\sigma_a(N_f)}{\tau_a(N_f)}. \tag{6.10}$$

Table 6.4. Parameters of the S–N curves in the local system for tested joints for R = −1

Welded joint	A	m	r	A_τ	m_τ	r_τ	k
Flange-tube (Sonsino)	17.034	4.306	0.965	25.782	8.233	0.974	1.65*
Tube-tube (Sonsino)	16.342	4.207	0.968	–	–	–	1.65*
Sonsino – total	15.015	3.658	0.940	25.782	8.233	0.974	1.36*
Flange-tube (Witt)	16.199	4.227	0.982	24.033	7.435	0.958	1.74**
Flange-tube (Küppers)	15.615	5.124	0.918	14.598	5.159	0.897	1.62**
Tube-tube (Küppers)	17.324	5.541	0.847	19.895	8.107	0.390	2.23**

*$N_f = 3 \cdot 10^5$ cycles
**$N_f = 5 \cdot 10^5$ cycles

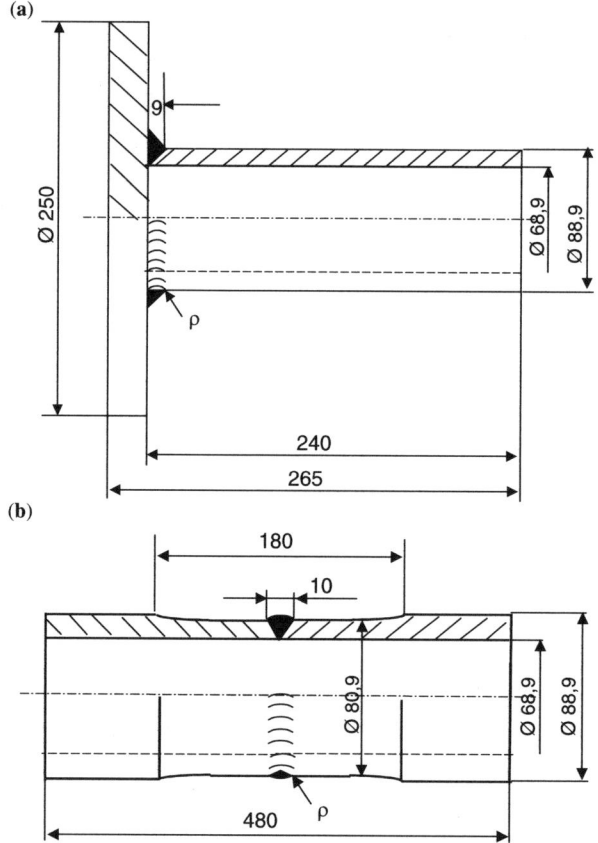

Fig. 6.1. Geometries of welded joints (**a**) flange-tube (FT), (**b**) tube-tube (TT)

All the experimental data for flange-tube and tube-tube specimens are shown in Figs. 6.2, 6.3, 6.4, 6.5, 6.6. According to the standards [38], a slope of the S–N curves m = 5, is typical for welded joints (see Sect. 1.1). As it can be seen in Table 6.4, under bending inclinations (m = 3.658–5.541) were obtained, and they were equal to the recommended inclination (m = 5). It was shown in the next figures as the reference scatter band for fatigue test results obtained under more complex loading. The obtained conformity shows whether the method of evaluation of the results obtained under complex loading is suitable for formulation the equivalent parameter under the fatigue damage.

Table 6.5 presents scatters of the fatigue test results related to the determined S–N curves for pure bending. In lines 1–5 there are scatters for particular fatigue tests. Line 6 presents the mean values calculated on the basis of scatters for all the tests. As for the Sonsino's tests, the total characteristic has been determined and it is shown in line 7. It can be seen that the mean scatter equal to 1 was obtained in all the cases. If theoretical notch coefficients based on the fictitious notch radius in the welded joint coincide with the fatigue one (line 8), then the mean scatters for flange-tube and tube-tube joints (line 9) are equal to 1 in relation to characteristics determined from the total characteristic of the considered welded joints. The mean standard deviation sT_N is 1.65, two standard deviations are $2sT_N = 3.331$. Let us notice a similarity with the often assumed scatter band band with the coefficient 3. For the significance level α = 5%, the mean scatter is $T_N = tsT_{N5\%/2} = 3.748$. Lower scatters can be observed for steel joints, and higher scatters are for aluminium joints. The scatters at the level $2sT_N$ are shown in Figs. 6.2, 6.3, 6.4, 6.5, 6.6, 6.7. All the calculation results are compared with those scatters.

From the calculation results presented in Table 6.5 it appears that in a complex case or under variable-amplitude loading the considered algorithm can be assumed as correct, if the calculation results are included into these scatter bands.

Table 6.5. Scatters of the fatigue test results related to S–N curves for bending

No	Welded joint	$\overline{T_N}$	sT_N	T_N (α=5%)	T_N ($2sT_N$)
1	Flange-tube (Sonsino)	1	1.641	3.878	3.282
2	Tube-tube (Sonsino)	1	1.455	3.101	2.910
3	Flange-tube (Witt)	1	1.492	3.172	2.992
4	Flange-tube (Küppers)	1	1.675	4.099	3.350
5	Tube-tube (Küppers)	1	2.061	4.491	4.121
6	All the joints together	1	1.665	3.748	3.331
7	Sonsino – total	1	1.758	3.687	3.565
8	Flange-tube (Sonsino) related to total	1.738	1.652	4.107	3.304
9	Tube-tube (Sonsino) related to total	1.282	1.521	2.732	3.042

76   6 An Example of Fatigue Life Evaluation Under Complex Loading States

Figures 6.7 and 6.8 show the weighed amplitudes of m degree for random tests of steel and aluminium welded joints [107, 157]. The amplitudes expressed by the following equation are most often applied by researchers

$$\sigma_{aw} = \left[\frac{\sum_{i=1}^{n} \sigma_{ai}^{m} n_i}{\sum_{i=1}^{n} n_i} \right]^{1/m}. \tag{6.11}$$

From Fig. 6.7 it appears that the weighed amplitudes for steel welded joints under bending are included into the scatter band for cyclic bending. From Fig. 6.8 it results that the weighed amplitudes for welded aluminium joints under bending are not included into the scatter bend for cyclic bending. Thus, the simple Palmgren-Miner rule is probably not valid in such a case.

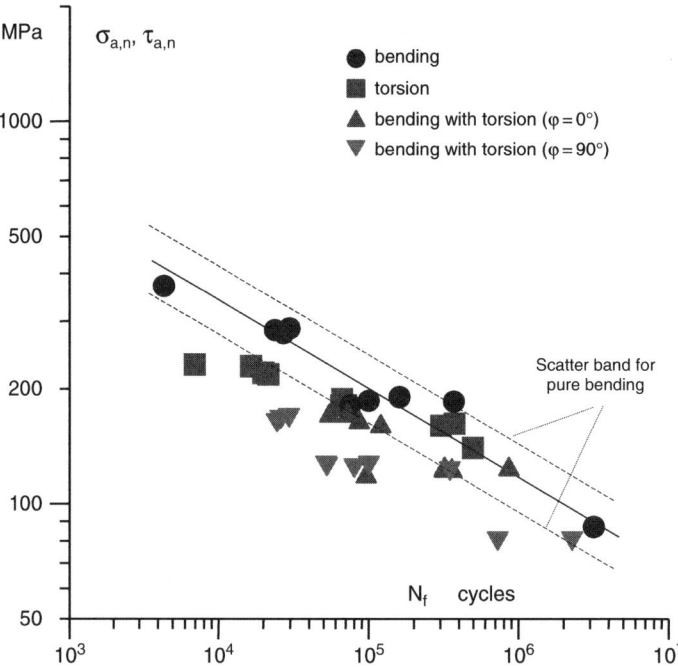

Fig. 6.2. Experimental data for flange-tube welded joints FT according to tests by Sonsino [217]

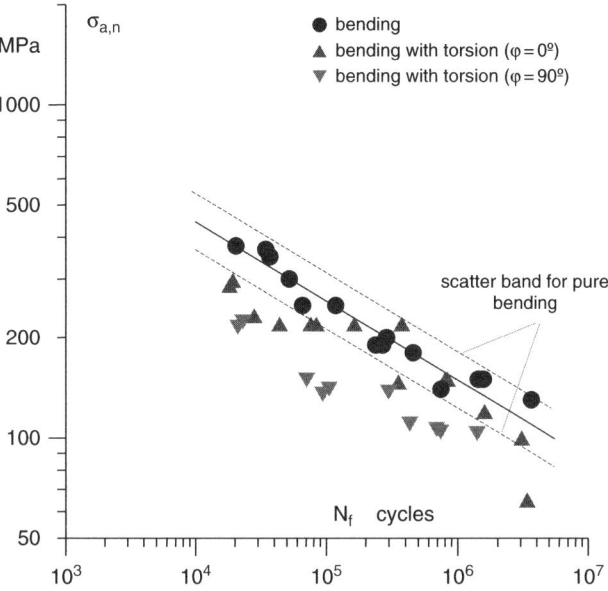

Fig. 6.3. Experimental data for tube-tube welded joints TT according to tests by Sonsino [217]

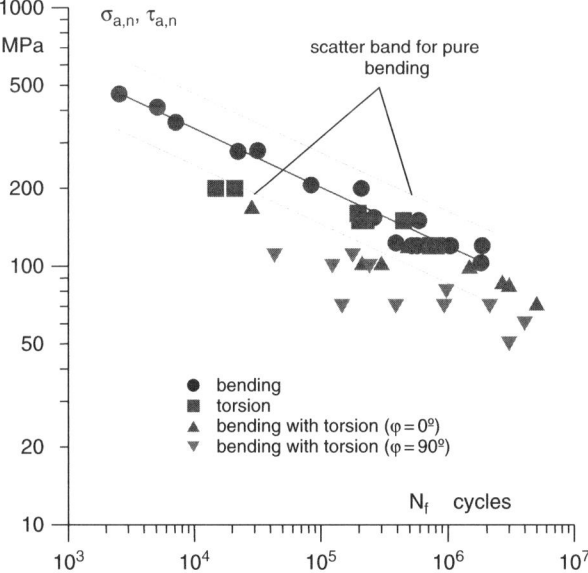

Fig. 6.4. Experimental data for flange-tube welded joints FT according to tests by Witt [240]

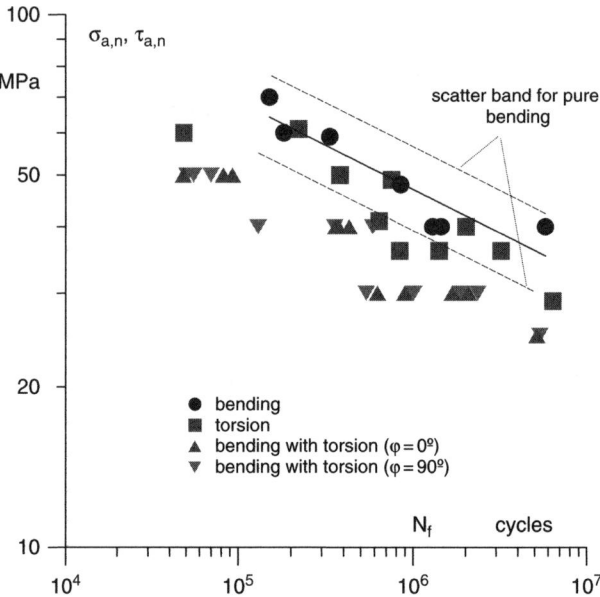

Fig. 6.5. Experimental data for flange-tube welded joints FT according to tests by Küppers [86, 88, 226]

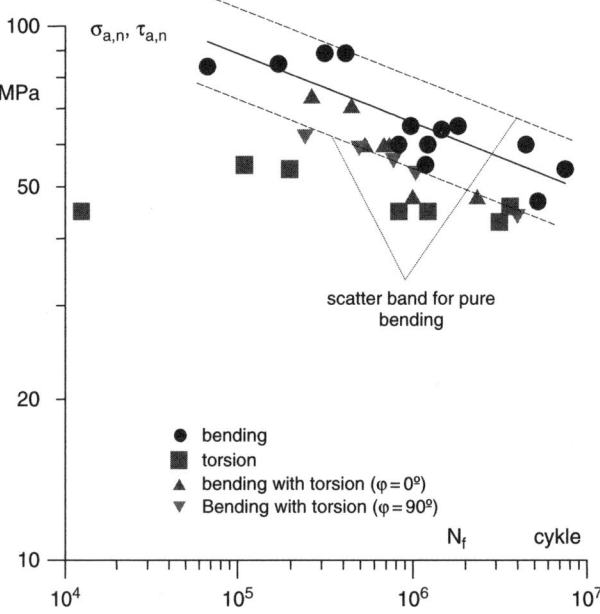

Fig. 6.6. Experimental data for tube-tube welded joints TT according to tests by Küppers [86, 88, 226]

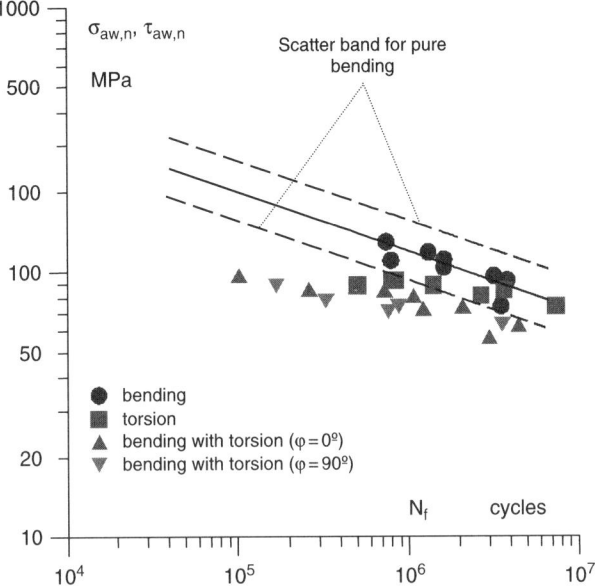

Fig. 6.7. Fatigue test results for steel welded joints under variable-amplitude loading according to Witt FT [104]

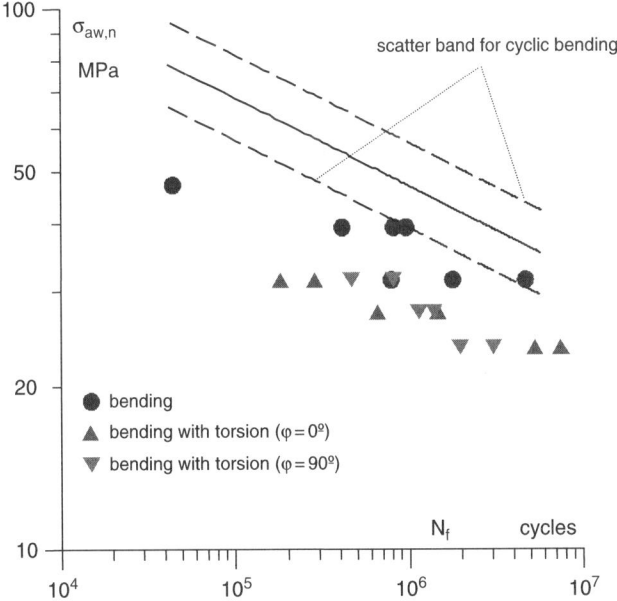

Fig. 6.8. Fatigue test results for aluminium welded joints under variable-amplitude loading according to Küppers FT [87]

Table 6.6. The applied abbreviations

Lp.	Type of loading	Calculations related to the S-N curve	Notation	Tests
1	Flange-tube	Flange-tube	FT	Sonsino Witt Küppers
2	Tube-tube	Tube-tube	TT	Sonsino Küppers
3	Flange-tube	Total characteristics	FTc	Sonsino
4	Tube-tube	Total characteristics	TTc	Sonsino

6.2 Verification of the Criteria Under Constant-amplitude Loading

The proposed criteria have been verified and the results of verification have been presented here. Some abbreviations have been introduced (Table 6.6).

6.2.1 The Parameter of Shear and Normal Strain Energy Density on the Critical Plane Determined by the Parameter of Normal Strain Energy Density

In the following subchapter the verification of the criterion assuming that the plane determined by maximum parameter of normal strain energy density in the critical plane is presented. Then, in this plane a sum of the normal strain energy density parameter (with the weight coefficient 1) and the shear strain energy density parameter in this plane is determined. As it was said, the weight (coefficient β) that should be assumed for the shear strain energy density parameter has not been determinated. Thus, analysis of the scatter $\overline{E}$ was performed for steel welded joints, determined according to (3.33) for all non-proportional tests and different values of the coefficient β. Values of scatters $\overline{E}$ versus coefficient β including the shear strain energy density parameter for steel welded joints are shown in Fig. 6.9. From this figure it appears that a value of the coefficient β varies within 7–14, depending on the tests. Its average value of 10 can be assumed.

6.2 Verification of the Criteria Under Constant-amplitude Loading

Figures 6.11–6.17 show the comparison of calculated and experimental lives for particular tests and for coefficients β determined from Fig. 6.9 including the shear strain energy density. It is important to note that for the tests of tube-flange joints performed by Sonsino (Fig. 6.10), the scatter is included into the band for pure bending except for torsion. When calculations are related to the total characteristics (Fig. 6.11), the calculated lives decrease and exceed the scatter band for cyclic tests. As for tube-tube welded joints, (Fig. 6.12), only one point for proportional and one point for non-proportional loading are outside the scatter band for pure bending. It is important to note that for tube-tube joints there are no tests under pure torsion because the cracks occur outside the joint in the native material. When calculations are related to the total fatigue characteristics (Fig. 6.13) similar results have been obtained. As for tests by Witt (Fig. 6.14), the tests results are similar as those obtained by Sonsino (see Fig. 6.10).

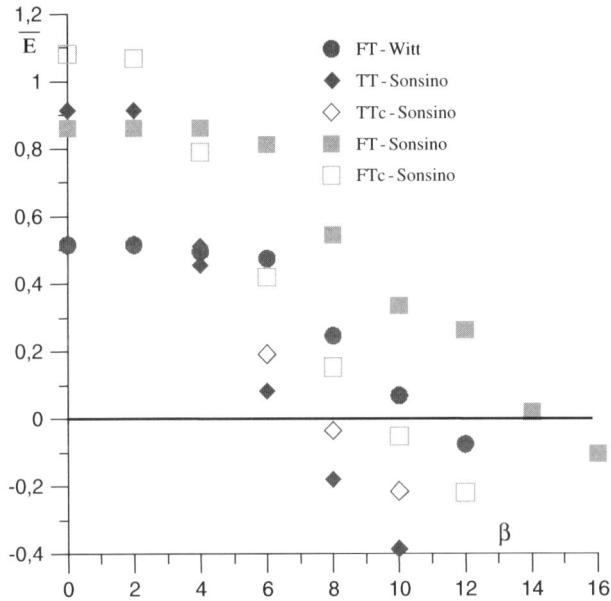

Fig. 6.9. Scatters versus shear strain energy density parameters for steel welded joints

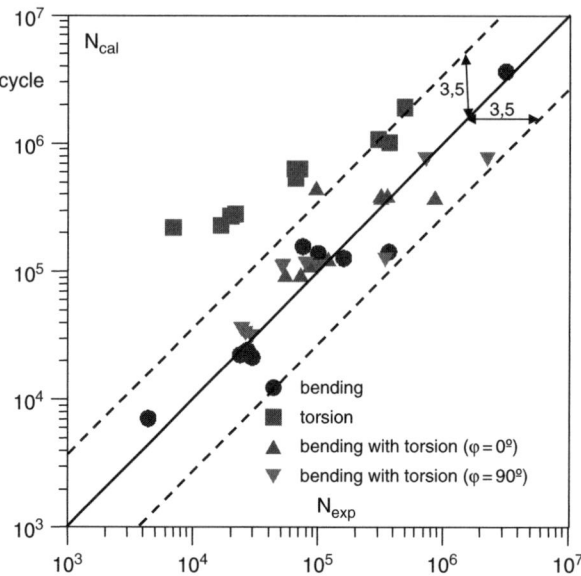

Fig. 6.10. Comparison of the calculated fatigue lives for flange-tube welded joints (FT) according to the criterion in the plane determined by the normal strain energy density parameter with the lives obtained by Sonsino

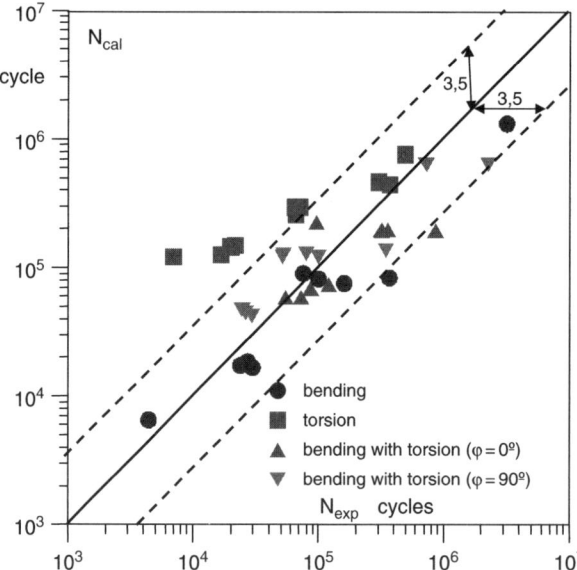

Fig. 6.11. Comparison of the calculated lives for flange-tube welded joints (FTc) according to the criterion in the plane determined by the normal strain energy density parameter with the lives obtained by Sonsino related to the total fatigue characteristics

6.2 Verification of the Criteria Under Constant-amplitude Loading 83

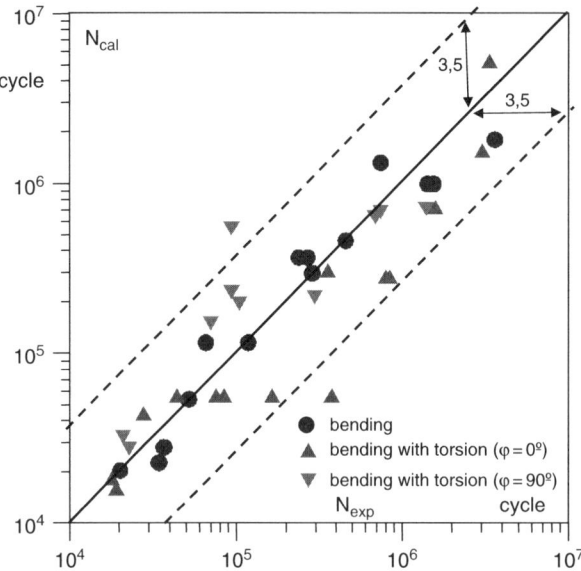

Fig. 6.12. Comparison of the calculated fatigue lives for tube-tube welded joints (TT) according to the criterion in the plane determined by the normal strain energy density parameter with the lives obtained by Sonsino

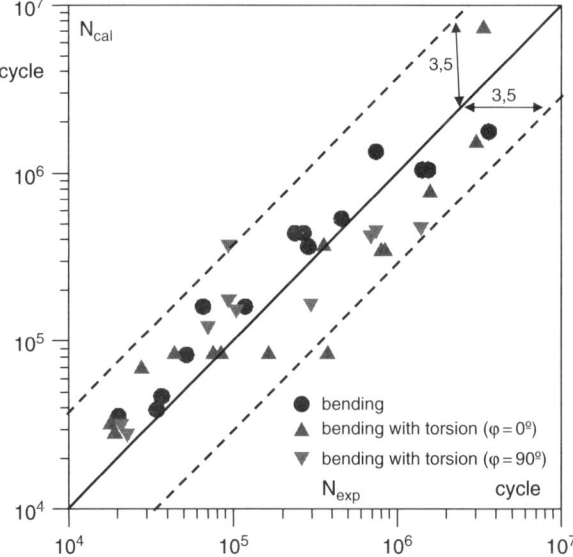

Fig. 6.13. Comparison of the calculated fatigue lives for tube-tube welded joints (TTc) according to the criterion in the plane determined by the normal strain energy density parameter with the lives obtained by Sonsino related to the total fatigue characteristics

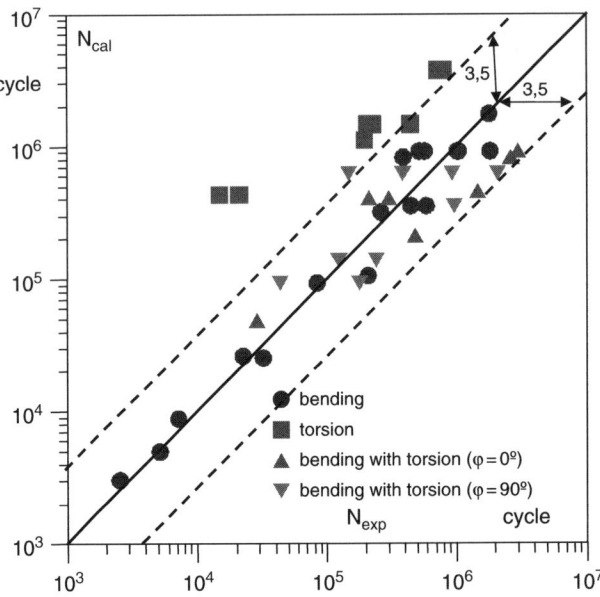

Fig. 6.14. Comparison of the calculated fatigue lives for flange-tube welded joints (FT) according to the criterion in the plane determined by the normal strain energy density parameter with the lives obtained by Witt for equal frequencies of bending with torsion

In the case of the tests of aluminium joints performed by Küppers, overestimated calculated fatigue lives were obtained for both tube-flange and tube-tube joints (Figs. 6.15 and 6.16). That overestimation can be even 100 times higher than the experimental value. During these tests, the values of the coefficient β, including participation of the shear strain energy density parameter in the critical plane, were not searched because favourable results had not been obtained for proportional loading.

6.2 Verification of the Criteria Under Constant-amplitude Loading   85

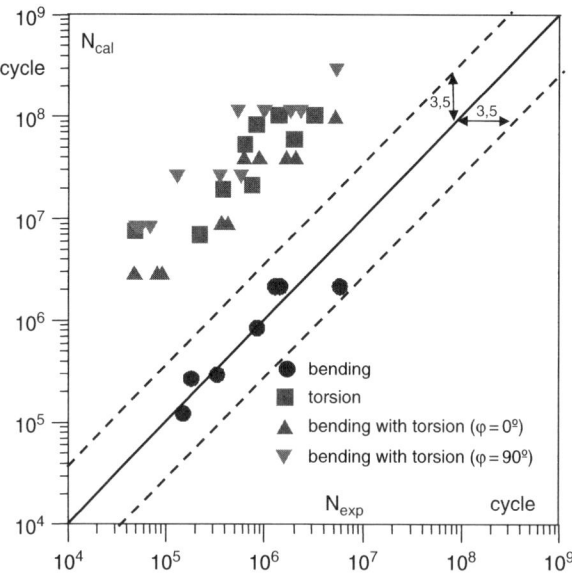

Fig. 6.15. Comparison of the calculated fatigue lives for flange-tube welded joints (FT) according to the criterion in the plane determined by the normal strain energy density parameter with the lives obtained by Küppers

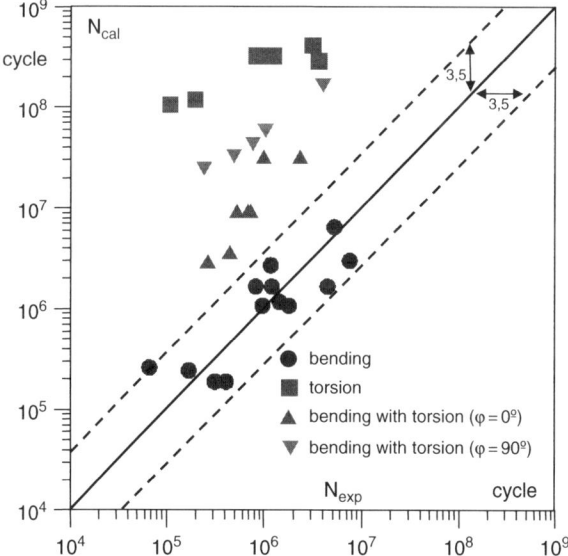

Fig. 6.16. Comparison of the calculated fatigue lives for tube-tube welded joints (TT) according to the criterion in the plane determined by the normal strain energy density parameter with the lives obtained by Küppers

6.2.2 The Parameter of Shear and Normal Strain Energy Density in the Critical Plane Determined by the Shear Strain Energy Density Parameter

In this part the criterion assuming the plane determined by the maximum value of shear strain energy density parameter as the critical plane was verified. In this plane, a sum of the normal strain energy density parameter and the shear strain energy density parameter is determined.

Figures 6.17–6.23 present comparisons of the calculated and experimental fatigue lives for particular tests. After the analysis of the experiment and the calculations it can be stated that in the case of the tests performed by Sonsino for tube-flange joints (Fig. 6.17) the results are included into the scatter band for pure bending, except for one point for proportional loading. If the calculation results are related to the total characteristics (Fig. 6.18), the calculated lives decrease like in the case when the plane of the normal strain energy density parameter was assumed as the critical plane (see Figs. 6.10 and 6.11). As for tube-tube welded joints (Fig. 6.19) it is possible to find that only one point for proportional loadings and one point for non-proportional loadings are located outside the scatter band for pure bending, like in the case when the plane determined by the normal strain energy density parameter is assumed as the critical plane (Fig. 6.12). Similar results were obtained for the tests related to the total fatigue characteristics (Fig. 6.20) and in Fig. 6.13. During the tests performed by Witt, when the plane of the shear strain energy density parameter was assumed as the critical plane (Fig. 6.21), a little poorer results were obtained as compared with the case when the plane determined by the normal strain energy density parameter was assumed as the critical plane (Fig. 6.14). It concerns, however, only the stress levels close the fatigue limit, where scatters of experimental results are greater. In the case of torsion, it can be stated that the calculated results from Fig. 6.21 are better than those in Fig. 6.14.

As for the tests by Küppers for aluminium welded joints, much better results were obtained than in the case when the plane of the normal strain energy density parameter was assumed as the critical plane for tube-flange joints (Fig. 6.22) and tube-tube joints (Fig. 6.23). Only some points are located outside the scatter band for pure bending.

6.2 Verification of the Criteria Under Constant-amplitude Loading 87

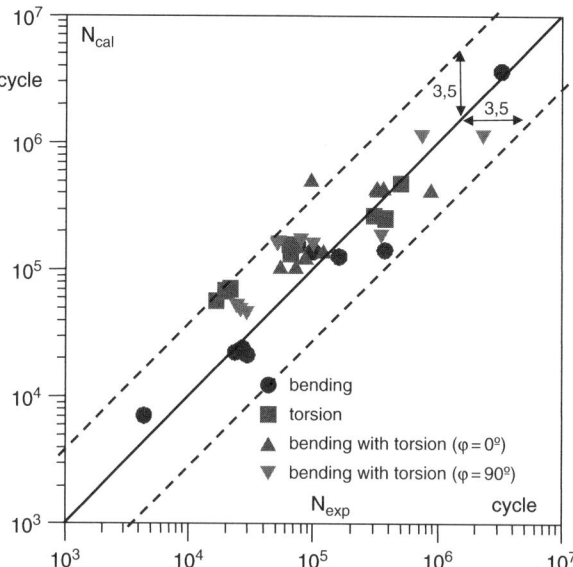

Fig. 6.17. Comparison of the calculated fatigue lives for flange-tube welded joints (FT) according to the criterion in the plane determined by the shear strain energy density parameter with the lives obtained by Sonsino

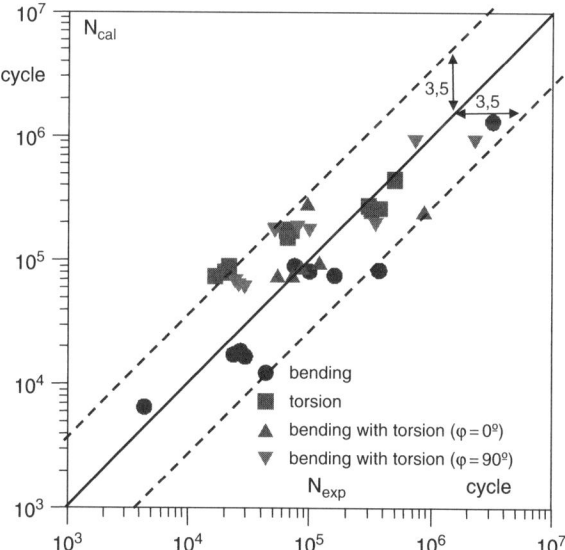

Fig. 6.18. Comparison of the calculated fatigue lives for flange-tube welded joints (FTc) according to the criterion in the plane determined by the shear strain energy density parameter with the lives obtained by Sonsino in relation to the total fatigue characteristics

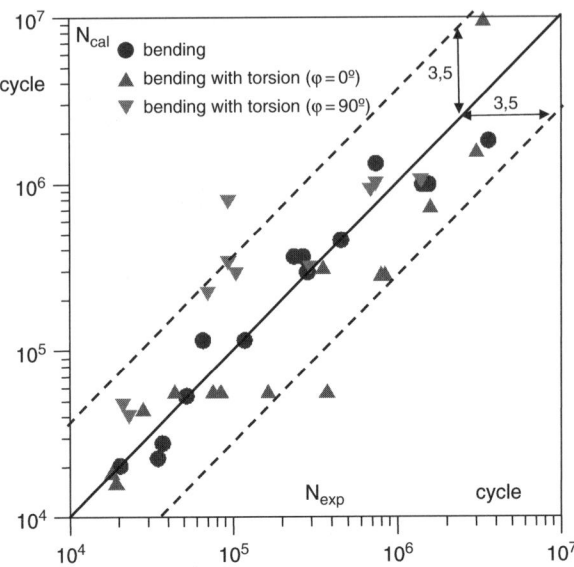

Fig. 6.19. Comparison of the calculated fatigue lives for tube-tube welded joints (TT) according to the criterion in the plane determined by the shear strain energy density parameter with the lives obtained by Sonsino

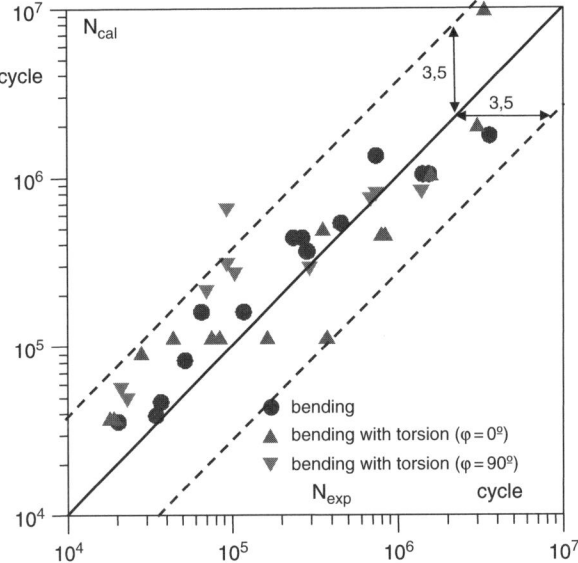

Fig. 6.20. Comparison of the calculated fatigue lives for tube-tube welded joints (TTc) according to the criterion in the plane determined by the shear strain energy density parameter with the lives obtained by Sonsino related to the total fatigue characteristics

6.2 Verification of the Criteria Under Constant-amplitude Loading 89

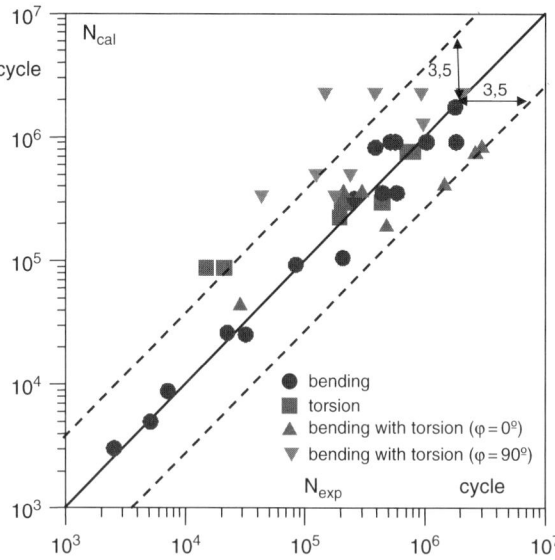

Fig. 6.21. Comparison of the calculated fatigue lives for flange-tube welded joints (FT) according to the criterion in the plane determined by the shear strain energy density parameter with the lives obtained by Witt under equal frequencies of bending and torsion

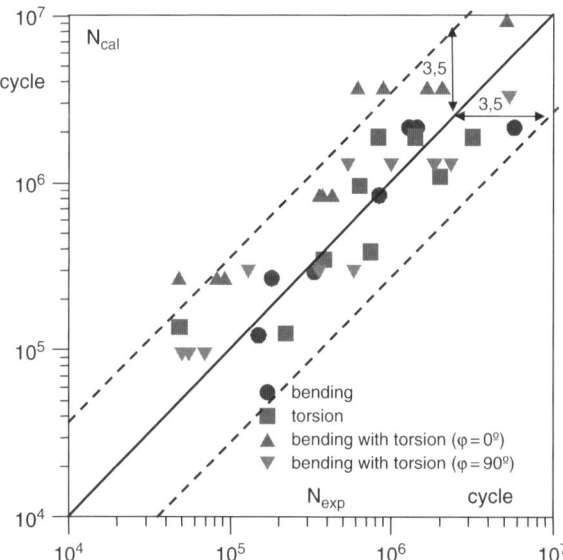

Fig. 6.22. Comparison of the calculated fatigue lives for flange-tube welded joints (FT) according to the criterion in the plane determined by the shear strain energy density parameter with the lives obtained by Küppers

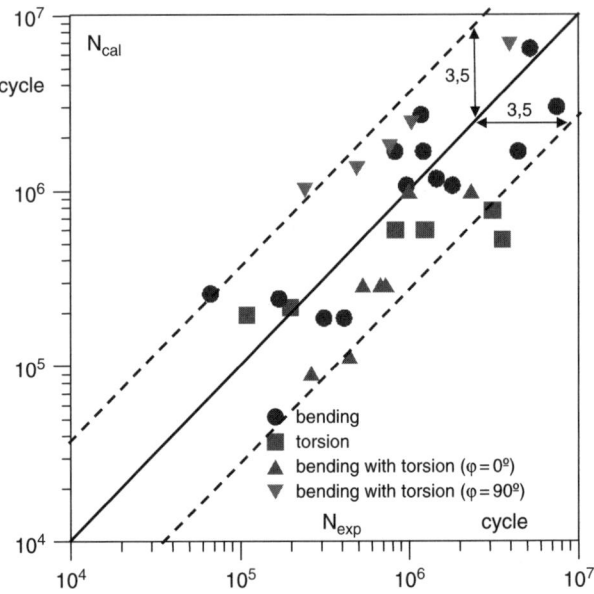

Fig. 6.23. Comparison of the calculated fatigue lives for tube-tube welded joints (TT) according to the criterion in the plane determined by the shear strain energy density parameter with the lives obtained by Küppers

A qualitative (not quantitative) description was used for analysis of the results shown in Figs. 6.10–6.23. Thus, according to the suggestions included into Sect. 3.2, the mean scatters $\overline{T}_N$ were determined for lives and the scatter bands T_N were determined under a double standard deviation. The calculation results are shown in Table 6.7. From the data presented in this table it appears that if the plane determined by the maximum shear strain energy density parameter is assumed as the critical plane, less mean scatters and scatter coefficients are obtained in comparison with the case when the plane determined by the maximum normal strain energy density parameter is understood as the critical plane. It is evident in the case of the tests by Küppers for welded aluminium joints. Only in one case (the tests by Sonsino for tube-tube welded joints) a contrary effect can be observed. However, the differences are small for steel welded joints for both considered critical planes. Thus, in this case one of two possible critical planes can be assumed: the plane determined by the normal strain energy density parameter (like for cast irons), or by the shear strain energy density parameter (like for steels). Here, it should be noted that the weld structure has a brittle material character, like for example, cast iron.

Table 6.7. Scatters of the cyclic test results according to the selected criteria

Critical plane		Sonsino				Witt	Küppers	
	FT	TT	FTc	TTc		FT	FT	TT
W_η $\overline{T}_N$	2.046	1.208	1.236	1.106		1.683	40.365*	64.565*
T_N	3.936	3.936	5.047	3.900		7.835	3.656*	10.790*
$W_{\eta s}$ $\overline{T}_N$	1.538	1.005	1.439	1.340		1.549	1.479	1.330
T_N	3,162	4,477	4,266	3,954		5,673	4,083	5,129

* only proportional loadings

If the plane determined by the maximum normal strain energy density parameter is assumed as the critical plane, after non-proportional tests, the value of the coefficient β including a part concerning the shear strain energy density parameter in the expression for the equivalent strain energy parameter should be determined. Thus, it is much better to apply the energy criterion based on the critical plane determined by the shear strain energy density parameter. This criterion was used for further calculations.

After the comparison of the scatter bands for uniaxial cyclic tests under pure bending (Table 6.5), when the mean scatter band varies about 3.5 and under combined bending with torsion with constant frequencies (Table 6.7) it can be seen that under complex loading the scatter band increase in relation to simple loadings, except for the tests by Sonsino for flange-tube (FT) welded joints.

6.2.3 The Influence of Different Frequencies of Bending and Torsion on Fatigue Life

Witt performed also tests under different frequencies of bending, f_σ and torsion, f_τ under the following combinations

$$f_\sigma = 5f_\tau \tag{6.12}$$

and

$$f_\sigma = 0.2f_\tau. \tag{6.13}$$

The calculated and experimental results for all the cyclic tests are compared in Fig. 6.24, where the assumed number of cycles comes from bending. For those tests, the mean scatter $\overline{T}_N = 1.306$ and the scatter band with coefficient $T_N = 4.667$, were obtained. The obtained scatter band is less than that for tests for equal frequencies (Table 6.7), and greater than that

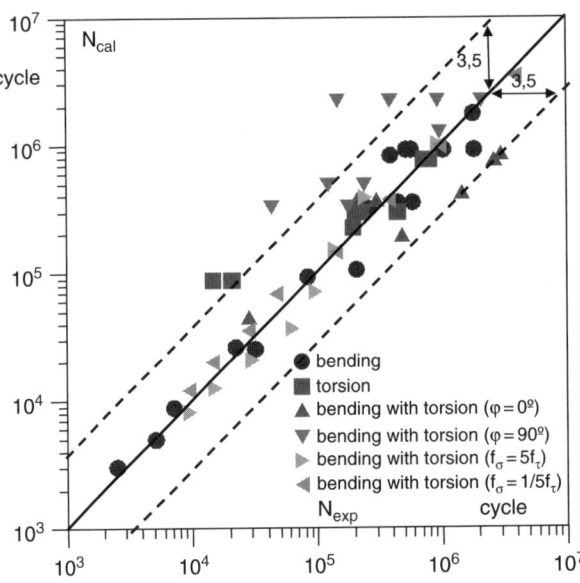

Fig. 6.24. Comparison of the calculated fatigue lives for flange-tube welded joints (FT) according to the criterion in the plane determined by the shear strain energy density parameter with the lives obtained by Witt under different frequencies of bending and torsion

for pure bending (Table 6.5). From the analysis of results it appears that for loadings with different frequencies the obtained conformity of calculation and the experiment was very good.

6.3 Verification Under Variable-amplitude Loading

The reported variable-amplitude tests concerned flange-tube welded joints according to Witt, and aluminium joints according to Küppers.

The results for steel welded joints are shown in Fig. 6.25. The weighed amplitude σ_{aw} according to (6.11) is expressed by

$$\sigma_{aw} = 0.370\, \sigma_{amax}. \tag{6.14}$$

The determined mean scatters are $\overline{T}_N = 1.268$, and the scatter band is $T_N = 3.846$. The scatters are greater than those for pure bending (Table 6.5), but lesser than those for combined bending with torsion (Table 6.7).

Good results of calculations were obtained as compared with the experimental data.

The results for welded aluminium joints are presented in Fig. 6.26. The weighed amplitude, σ_{aw} according to (6.11), is expressed as

$$\sigma_{aw} = 0.393\, \sigma_{amax}. \qquad (6.15)$$

The calculated results for pure bending are overestimated in relation to the experimental results. While searching a description of this phenomenon, the correction coefficient was applied, like in the Serensen-Kogayev hypothesis. The correction coefficient can be expressed as

$$b' = \frac{\sigma_{aw}}{\sigma_{a\,max}}. \qquad (6.16)$$

According to (6.15) and (6.16), b'= 0.393, which means that it is less than one and close to 1/3. Next calculations of life were performed according to

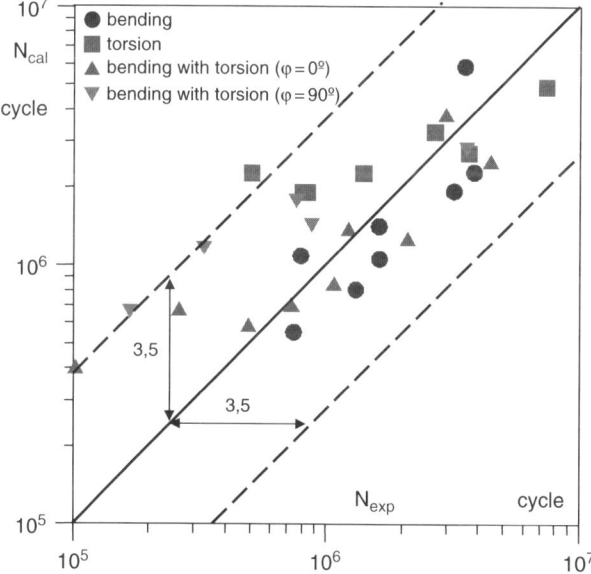

Fig. 6.25. Comparison of the calculated and experimental fatigue lives for flange-tube welded joints (FT) under variable-amplitude loading according to the criterion in the plane determined by the shear strain energy density parameter with the lives obtained by Witt

that modification of the Palmgren-Miner hypothesis. The determined mean scatters were $\overline{T}_N = 1/0.940 = 1.063$, and the scatter band was $T_N = 4.519$, and in this case the scatters are less than those for tests under pure bending (Table 6.5) and under combined bending with torsion (Table 6.7). The differences are not very high, so it can be assumed that the obtained results are satisfactory.

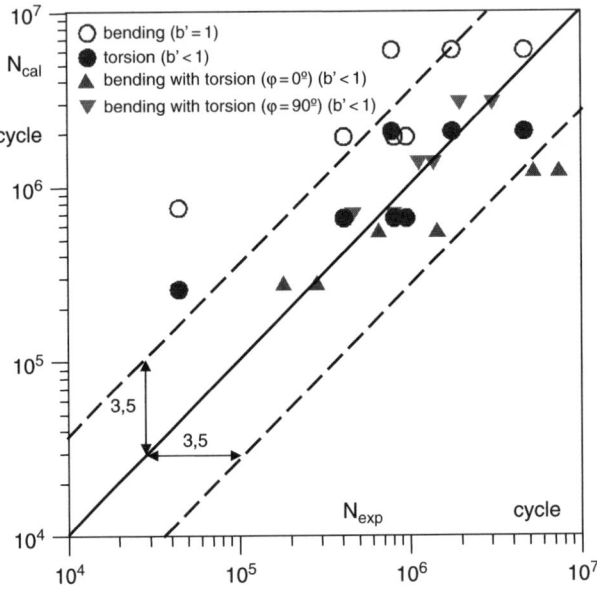

Fig. 6.26. Comparison of the calculated fatigue lives under variable-amplitude loadings for flange-tube welded joints (FT) according to the criterion in the plane determined by the shear strain energy density parameter with the lives obtained by Küppers

7 Conclusions

On the basis of the performed analyses and calculations the following conclusions can be drawn:
1. In the case of tests under uniaxial cyclic loading it can be stated that:
   1.1 It is possible to determine one common fatigue characteristics for four considered materials.
   1.2 Fatigue strength does not strongly depend on a type of the considered welded joint; much greater is the influence of loading (bending and axial loading)
   1.3 The calculated sum of damages for normal distribution (Gaussian spectrum) and normal distribution with overloads are included into the scatter band with coefficient 3 according to the Palmgren-Miner hypothesis and with the significance level 5%.
   1.4 In the considered joints, the mean stress values do not influence fatigue life; the obtained fatigue curves for symmetric and pulsating loadings were almost identical. It probably results from the existing high residual stresses, previously measured.

2. In the case of multiaxial loading it can be concluded that:
   2.1 Before estimation of multiaxial fatigue history in welded joints by local stresses and strains, the actual local radius at the weld edge should be determined. Owing to the fictitious local radius, what in the worst case for sharp notches $\rho=0$ means a crack, the notch coefficients for bending, K_{fb} and for torsion, K_{ft} can be calculated. Therefore, it is necessary to define the fictitious notch radii ρ_f for bending and torsion separately. In the case of welded steel joints, the radii are $\rho_{fb} = 1.16$ mm for bending and $\rho_{ft} = 0.4$ mm for torsion.
   2.2 The normal and shear strain energy density parameters in the critical plane determined by the parameter of shear and normal strain energy density for steel welded joints give comparable results. If the plane determined by normal strain energy density parameter is assumed as the critical plane, it is necessary to define the experimental weight function including the shear strain energy density parameter in this plane. Thus, the application of

the energy criterion defined in the plane determined by the shear strain energy density parameter is recommended.

2.3 In the case of welded aluminium joints, satisfactory results of fatigue life calculations were obtained for the criterion of energy parameter in the plane defined by the shear strain energy density parameter. When the energy criterion was applied in the plane defined by the normal strain energy density parameter, the calculated results were overestimated in comparison with the experimental ones.

2.4 Application of the maximum shear and normal strain energy density parameter in the critical plane for aluminium welded joints subjected to variable-amplitude bending with torsion is right if the Palmgren-Miner hypothesis is taken into account, and the correction coefficient is used, like in case of the Serensen-Kogayev hypothesis. The correction coefficient is the quotient of the weighed amplitude of the slope degree of the fatigue curve and the maximum amplitude.

References

1. Achtelik H., Lachowicz C., Łagoda T., Macha E.: Fatigue characteristics of the notched specimens of 10HNAP steel under cyclic and random synchronous bending with torsion, presented in 2nd annual meeting of Copernicus Contract CIPA CT940194, Metz1996, L. Toth, ed., Miscolc – 1997, pp. 176–197
2. Achtelik H., Lachowicz C., Łagoda T., Macha E.: Life time of the notched specimens of 10HNAP steel under proportional bending with torsion, in Proceedings and presented in 1st annual fatigue group meeting of Copernicus Contract CIPA CT940194, Smolenice 1996, M. Bily, ed., Slovak Academy of Science – 1997, pp. 60–69
3. Achtelik H., Łagoda T., Macha E.: Trwałość zmęczeniowa elementów gładkich i z karbem poddanych cyklicznym obciążeniom giętnoskrętnym, XI Konferencja Naukowa – Problemy Rozwoju Maszyn Roboczych, Zakopane 1998, Vol. II, ss. 15–22
4. Adib H., Gilgert J., Pluvinage G.: Fatigue life duration prediction for welded prediction for welded spots by Volumetric method, Int. J. Fatigue, Vol. 26, 2004, pp. 81–94
5. Agerskov H.: Fatigue in steel structures under random loading, J. Constructional Steel Research, Vol. 53, 2000, pp. 283–305
6. Agerskov H., Ibsø J.B.: Fatigue life of plate elements with welded transverse attachments subjected to stochastic loading, A.F. Blom, ed., Proceedings of Nordic Conference on Fatigue, West Midlands, UK, EMAS Publishers, 1993
7. Amzalang C., Gerey J.P., Robert J.L., Bahuaud J.: Standarization of the rainflow counting method for fatigue analysis, Int. J. Fatigue, Vol. 17, 1994, pp. 287–293
8. Anthes R.J., Kottgen V.B., Seeger T.: Kerbformzahlen von Stumpfstossen und Doppel-T-Stossen Mitteilung aus dem Fachgebiet Werkstoffmechanik der Techn, Hochschule Darmstadt Schewissen und Schneiden 45, 1993, Heft 12, ss. 685–688
9. ASTM E 739-91 (1998), Standard Practice for Statistical Analysis of Linearized Stress-Life (S – N) and Strain-Life (ε – N) Fatigue Data. In: Annual Book of ASTM Standards, Vol. 03.01, Philadelphia 1999, pp. 710–718
10. ASTM E 1049-85, Standard practices for cycle counting in fatigue analysis. In: Annual Book of ASTM Standards, Vol. 03.01, Philadelphia 1999, pp. 614–620

11. Atzori B., Meneghettil G., Susmel L.: Multiaxial fatigue life predictions by using modified Manson-Coffin curves, In: Proceedings of the 7th Internat. Conference on Biaxial/Multiaxial Fatigue and Fracture, Berlin 2004, DVM, C.M. Sonsino, H. Zenner, P.D. Portella, eds., pp. 129–134
12. Bäckström M., Marquis G.: Interaction equations for multiaxial fatigue assessment of welded structures, Fatigue Fract. Engng Mater. Struct., Vol. 27, 2004, pp. 991–1003
13. Banvillet A., Łagoda T., Macha E., Niesłony A., Palin-Luc T., Vittori J.F.: Fatigue life under non-gaussian random loading from various models, Int. J. Fatigue, Vol. 26, No. 4, 2004, pp. 349–363
14. Banvillet A., Łagoda T., Macha E., Niesłony A., Palin-Luc T., Vittori J.F.: Metody oceny trwałości zmęczeniowej w losowym stanie obciążenia na przykładzie badań stali 10HNAP, XXI Sympozjon PKM, Ustroń 2003, WNT, red. J. Wojnarowski i J. Drewniak, Tom I, 2003, ss. 151–160
15. Banevillet A., Palin-Luc T., Vittori J.F., Łagoda T., Macha E.: Evaluation of critical plane approaches in predicting long fatigue life of a spheroidal graphite cast iron under non-proportional variable amplitude bending, proceedings of 6th ICBMFF, Lisbon 2001, M. M. de Freitas, ed., Vol. I, pp. 407–414
16. Basquin O.H.: The experimental law of endurance test, Proceedings of ASTM, Philadelphia, Vol. 10, 1910, pp. 625–630
17. Będkowski W., Gasiak G., Lachowicz C., Lampard D., Lichtarowicz A., Łagoda T., Macha E.: Zależność między erozją kawitacyjną a zmęczeniem losowym materiałów, Problemy Maszyn Roboczych, 1997, Vol. 9, ss. 107–123
18. Będkowski W., Gasiak G., Lachowicz C., Lichtarowicz A., Łagoda T., Macha E.: Relation between cavitation erosion resistance of materials and their fatigue strength under random loading, Wear, Vol. 230, 1999, pp. 201–209
19. Bellet D., Taylor D., Marco S., Mazzeo E., Guillois J., Pircher T.: The fatigue behaviour of three-dimensional stress concentrations, Int. J. Fatigue, Vol. 27, 2005, pp. 207–221
20. Bentachfine S., Pluvinage G., Gilgert J., Azari Z., Bouami D.: Notch effect in low cycle fatigue, Int. J. Fatigue, Vol. 21, 1999, pp. 421–430
21. Berger C., Eulitz K.G., Heuler P., Kotte K.L., Naundorf H., Schuetz W., Sonsino C.M., Wimmer A., Zenner H.: Betriebsfestigkeit in Germany – an overview, Int. J. Fatigue, Vol. 24, 2002, pp. 603–625
22. Bergman J., Seeger T.: On the influence of cyclic stress-strain curves, damage parameters and various evaluation concepts on the life prediction by the local approach. Proc. 2nd Europan Coll. On Fracture, Darmstadt, Germany, VDI-Report of Progress, Vol. 18, No.6, 1979
23. Bochenek A.: Elementy mechaniki pękania, Politechnika Częstochowska, Częstochowa, 1998

24. Bomas H., Linkiewitz T., Mayer P.: Fatigue limit prediction of notched hard steel parts under uniaxial and torsional loading, in: Engineering Against Fatigue, J.H. Beynon, M.W. Brown, R.A. Smith, T.C. Lindley and B. Tomkins, eds., Sheffild, Balkema 1999, pp. 349–356
25. Borri C.: Generation procedures of stationary random processes simulating wind time series, lecture in Aerodynamics in Struct. Engineering, The Ruhr University of Bochum, 1988, p. 35
26. Borgman L.E.: Ocean wave simulation for engineering design, J. of Waterways and Harbours Division, Vol. 95, No. WW4, 1969, pp. 557–583
27. Braccesi C., Cianetti F., Lori G., Pioli D.: Fatigue behaviour analysis of mechanical components subject to random bimodal stress process: frequency domain approach, Int. J. Fatigue, Vol. 27, 2005, pp. 335–345
28. Chen X., An K., Kim K.S.: Low-cycle fatigue of 1Cr-18Ni-9Ti stainless steel and related weld metal under axial, torsional and 90° out-of-phase loading, Fatigue Fract. Engng Mater. Struct., Vol. 27, 2004, pp. 439–448
29. Cheng G., Kuang Z.B., Lou Z.W., Li H.: Experimental investigation of fatigue behaviour for welded joint with mechanical heterogeneity, Int. J. Pres. Ves. Piping, Vol. 67, 1996, pp. 229–242
30. Corten H.T., Dolan T.J.: Cumulative fatigue damage, Presented at Int. Conference on fatigue of metals, Published by the Institution of Mechanical Engineers, London 1956, p. 15
31. Dang Van K., Bignonnet A., Fayard J.L.: Assessment of welded structures by a structural multiaxial fatigue approach, in: Biaxial/Multiaxial Fatigue Fracture, A. Carpinteri, M. de Freitas, A. Spagnoli, eds., Elsevier 2003, pp. 3–21
32. Dang Van K., Griveau B., Message O.: On a New Multiaxial Fatigue Criterion: Theory and Application. In: Biaxial and Multiaxial Fatigue, M.W. Brown and K.J. Miller, eds., EGF3, London 1989, Mechanical Enginering Publication, pp. 479–498
33. Dietrich M.: (red.), Podstawy konstrukcji Maszyn, Tom 2, WNT, Warszawa 1995, s. 656
34. Dowling N.E.: Fatigue failure prediction for complicated stress-strain histories, J. Mater., ASTM, Vol. 7, 1972, pp. 71–87
35. Downing S.D., Socie D.F.: Simple rainflow counting algorithms, Int. J. Fatigue, Vol. 5, 1982, pp. 31–44
36. Eibel M., Sonsino C.M., Kaufmann H., Zhang G.: Fatigue assessment of laser welded thin sheet aluminium, Int. J. Fatigue, Vol. 25, 2003, pp. 719–731
37. Ellyin F.: Fatigue damage, crack growth and life prediction, Chapman and Hall, 1996 p. 462
38. Eurocode 3: Design of Steel Structures – Part 1-1: General Rules for Buildings. European Committee for Standardisation, Brussels 1992, ENV 1993-1-1
39. Eurocode 9: Design of Aluminium Structures – Part 2: Structures susceptible to fatigue, Brussels 1998, ENV 1999–2
40. Fricke W.: Fatigue analysis of welded joints: state of development, Marine Structures, Vol. 16, 2003, pp. 185–200

41. Gasiak G.: Wybrane techniki pomiarowe, Skrypt Politechniki Opolskiej Nr 222, Opole 1999, s. 219
42. Gąska D.: Zastosowanie tensometrii oporowej do wyznaczania naprężeń w spoinach konstrukcji stalowych, Praca Magisterska, Politechnika Śląska, Katowice 2003, s. 89
43. Gołoś K.: Komputerowo wspomagana analiza widma obciążenia, Przegląd Mechaniczny, Z.2, 2001, ss. 7–10
44. Goss Cz., Kłysz S., Wojnowski W.: Problemy niskocyklowej trwałości zmęczeniowej wybranych stali i połączeń spawanych, Wydawnictwo Instytutu Technicznego Wojsk Lotniczych, Warszawa 2004, s. 154
45. Gurney T.R.: Zmęczenie konstrukcji spawanych, WNT, Warszawa 1973, s. 355
46. Haibach E.: Modifizierte Lineare Schadensakkumulations Hypothese zur Berucksichtigung des Dauerfestigkeitsabfalls mit fortschreitender Dchuadiggung, Technische Mitteilungen Nr 50/70, LBF, Darmstadt 1970, s. 15
47. Hobbacher A.: Recommendations for Fatigue Design of Welded Joints and Components. ISO Standard Proposal. IIW document XIII-1539-96/XV-845-96. International Institute of Welding, Paris, 1996
48. Hoffman H., Seeger T.: Stress strain analysis and life predictions of a notched shaft under multiaxial loading, Multiaxial Fatigue: Analysis and Experiments, AE-14, eds., G.E., Leese and D. Socie, Society of Automative Engineers, Inc., Werrendale, USA, 1989, pp. 81–99
49. Huck M., Schultz W., Fischer R., Kobler G.: A standard random load sequence of Gaussian type recommended for general application in fatigue testing, LBF-Report No 2909, IABG-Report No TF-570, April 1976, p. 21
50. Huo L., Wang D., Zhang Y.: Investigation of the fatigue behaviour of the welded joints treated by TIG dressing and ultrasonic peening under variable-amplitude, Int. J. Fatigue, Vol. 27, 2005, pp. 95–101
51. Huther I., Primot L., Lieurade J.J., Colchen D., Debicz S.: Weld quality and the cyclic fatigue strength of steel welded joints, Welding in the World, Vol. 35, 1995, pp. 118–133
52. Ibsø J.B., Agerskov H.: Fatigue life of offshore steel structures under stochastic loading, Report No. R 299, Department of Structural Engineering and Materials Technical University of Denmark, Lyngby, Danmark, 1992
53. Jakubczak H.: Niepewność danych w prognozowaniu trwałości zmęczeniowej konstrukcji nośnych maszyn, Prace naukowe, Mechanika, Z.194, Politechnika Warszawska, Warszawa 2002, s. 127
54. Jiang Y., Xu B.: Deformation analysis of notched components and assessment of approximate methods, Fatigue Fract. Engng Mater. Struc., Vol. 24, 2001, pp. 729–740
55. Kardas D., Kluger K., Łagoda T., Ogonowski P.: Fatigue life of AlCu4Mg1 aluminium alloy under constant amplitude bending with torsion, Proceedings of the Seventh International Conference on Biaxial/Multiaxial Fatigue and Fracture, DVM, Berlin 2004, pp. 197–202

56. Kardas D., Kluger K., Łagoda T., Ogonowski P.: Trwałość zmęczeniowa duraluminium PA6 w warunkach proporcjonalnego stałoamplitudowego zginania ze skręcaniem, Zmęczenie i Mechanika Pękania, Wyd. ATR w Bydgoszczy, Bydgoszcz 2004, ss. 163–169
57. Kardas D., Kohut M., Łagoda T.: Kumulacja uszkodzeń zmęczeniowych materiałów nieżelaznych w prostych stanach naprężenia, III Sympozjum Mechaniki Zniszczenia Materiałów i Konstrukcji, Augustów, 1 – 4.06.2005, Politechnika Białostocka, ss. 125–128
58. Kardas D., Łagoda T., Macha E., Niesłony A.: Fatigue life under variable-amplitude tension-compression according to the cycle counting and spectral methods, in: Fatigue Damage Cumulative, Seville, may 2003, Department of Mechanical and Materials Engineering University of Seville 2003, pp. 359–360
59. Kardas D., Łagoda T., Macha E., Niesłony A.: Lifetime under variable-amplitude tension-compression calculated in time and frequency domains using the strain energy density parameter, LCF 5, Berlin 3003, Abstracts, DVM, A63, ps 2
60. Kardas D., Łagoda T., Macha E., Niesłony A.: Porównanie kumulacji uszkodzeń zmęczeniowych według parametru odkształceniowego metodą zliczania cykli i metodą spektralną, IX Krajowa Konferencja Mechaniki Pękania. Kielce 2003, Politechnika Świętokrzyska, ss. 237–243
61. Karolczuk A.: Identyfikacja płaszczyzn złomu zmęczeniowego stali 18G2A metodą funkcji wagowych, ZN 292, Mechanika, z.77, Politechnika Opolska, Opole 2003, ss. 33–40
62. Karolczuk A.: Płaszczyzny krytyczne w modelach zmęczenie materiałów przy wieloosiowych obciążeniach losowych, Praca doktorska, Wydział Mechaniczny Politechniki Opolskiej. Opole 2003, s. 146
63. Karolczuk A., Łagoda T.: Położenie płaszczyzn krytycznych i złomów zmęczeniowych według parametru gęstości energii odkształceń, Problemy Maszyn Roboczych, Z.20, 2002, ss. 99–107
64. Karolczuk A., Łagoda T.: Zastosowanie kinematycznego modelu umocnienia przy obliczaniu trwałości zmęczeniowej w prostym stanie obciążenia, XX Sympozjon Podstaw Konstrukcji Maszyn, Zeszyty Naukowe Politechniki Opolskiej, Z.68, Nr 270/2001, Opole 2001, ss. 465–472
65. Karolczuk A., Łagoda T., Macha E.: Determination of fatigue life of 10HNAP and 1208.3 steels with the parameter of strain energy density, Proceedings of the Eight International Fatigue Congress, Fatigue 2002, Stockholm, Sweden, ed., A.F. Blom, EMAS, 2002, Vol. 1, pp. 515–522
66. Karolczuk A., Łagoda T., Macha E.: Wyznaczanie trwałości zmęczeniowej stali 1208.3 za pomocą parametru gęstości energii odkształceń w zakresie małej liczby cykli, Sympozjum Mechaniki Zniszczenia Materiałów i Konstrukcji, Augustów 2001, Zeszyty Naukowe Politechniki Białostockiej, Nauki Techniczne Nr 138, Mechanika Zeszyt 24, Białystok 2001, ss. 231–238

67. Karolczuk A., Łagoda T., Niesłony A.: Critical planes in the specimen section taking the stress gradient into account, Proceedings of 14th International Conference on Fracture, ECF 14, 8–12 September 2002, Kraków, Vol. 2, pp. 109–116
68. Karolczuk A., Łagoda T., Niesłony A.: Położenie płaszczyzn krytycznych w przekrojach elementów z uwzględnieniem gradientów naprężeń, Problemy Maszyn Roboczych, Z.18, 2001, ss. 51–60
69. Karolczuk A., Macha E.: Wyznaczanie płaszczyzn krytycznych metodami funkcji wagowych I kumulacji uszkodzeń przy wieloosiowych obciążeniach losowych, Zeszyty Naukowe Politechniki Świętokrzyskiej, Mechanika 78, 2003, ss. 245–252
70. Karolczuk A., Macha E.: Identification of fatigue fracture plane positions with the expected principal stress direction, International Conference on "Fatigue Crack Paths" (FCP 2003), ESIS, (CD), p. 8
71. Karolczuk A., Macha E.: Płaszczyzny krytyczne w modelach wieloosiowego zmęczenia materiałów, Wieloosiowe zmęczenie losowe maszyn i konstrukcji – Część VI, Studia i Monografie, Z.162, Politechnika Opolska, Opole 2004, s. 256
72. Karolczuk A., Macha E.: Położenia płaszczyzn złomu zmęczeniowego stali 18G2A i ich estymacja metodą funkcji wagowych, II Sympozjum Mechaniki Zniszczenia Materiałów i Konstrukcji, Politechnika Białostocka, Białystok 2003, ss. 147–151
73. Klimpel A., Dziubiński J.: Podstawy konstrukcji spawanych, Politechnika Śląska Skrypt Nr 507/21, Gliwice 1976, s. 195
74. Kluger K., Łagoda T.: Application of the Dang-Van criterion for life determination under uniaxial random tension-compression with different mean value, Fatigue Fract. Engng Mater. Struct., Vol. 27, 2004, pp. 505–512
75. Kluger K., Łagoda T.: Trwałość zmęczeniowa stali 10HNAP przy losowym rozciąganiu-ściskaniu z wartością średnią według kryterium Dang-Vana, Problemy Maszyn Roboczych, Z.24, 2004, ss. 29–46
76. Kluger K., Łagoda T.: Trwałość zmęczeniowa stali 10HNAP w warunkach jednoosiowego stałoamplitudowego i losowego obciążenia z różnymi wartościami średnimi, IX Krajowa Konferencja Mechaniki Pękania. Kielce 2003, Politechnika Świętokrzyska, ss. 253–260
77. Kluger K., Łagoda T.: Wpływ wartości średniej dla duraluminium PA6 w warunkach kombinacji zginania ze skręcaniem według parametru gęstości energii odkształceń, III Sympozjum Mechaniki Zniszczenia Materiałów i Konstrukcji, Augustów, 1 – 4 czerwca 2005, Politechnika Białostocka, ss. 145–148
78. Kluger K., Karolczuk A., Łagoda T., Macha E.: Application of the energy parameter for fatigue life estimation under uniaxial random loading with the mean value, ICF Turyn 2005, 20–25 March 2005, ed., A. Carpinteri, Politecnico di Torino, CD, ps 6

79. Kluger K., Łagoda T., Karolczuk A., Macha E.: Obliczanie trwałości zmęczeniowej przy obciążeniach eksploatacyjnych z wartością średnią za pomocą parametru energetycznego, Czasopismo Techniczne – Mechanika z.1-M/2005, Wydawnictwo Politechniki Krakowskiej, Kraków 2005, ss. 211–218
80. Kocańda S., Szala J.: Podstawy obliczeń zmęczeniowych, PWN, Warszawa 1985, s. 276
81. Kohout J., Věchet S.: A new function for fatigue curves characterization and its multiple merits, Int. J. Fatigue, Vol. 23, 2001, pp. 175–183
82. Kohut M., Łagoda T.: Badania zmęczeniowe mosiądzu MO58 w warunkach proporcjonalnego cyklicznego zginania ze skręcaniem, III Sympozjum Mechaniki Zniszczenia Materiałów i Konstrukcji, Augustów, 1 – 4 czerwca 2005, Politechnika Białostocka, ss. 159–162
83. Kohut M., Łagoda T.: Trwałość zmęczeniowa próbek okrągłych i kwadratowych ze stali 18G2A poddanych zginaniu wahadłowemu, Zmęczenie i Mechanika Pękania, Wyd. ATR w Bydgoszczy, Bydgoszcz 2004, ss. 195–202
84. Kowalczyk M.: Analiza wytrzymałościowa w procesie kształtowania spawanych konstrukcji nośnych maszyn górnictwa odkrywkowego, Transport Przemysłowy, No. 2(8), 2002, ss. 55–58
85. Kozicki I., Macha E.: Generowanie wzajemnie opóźnionych sygnałów losowych o żądanych charakterystykach statystycznych, Studia i Monografie z.22, Wyższa Szkoła Inżynierska, Opole 1988, s. 86
86. Kueppers M., Sonsino C.M.: Critical plane approach for the assessment of the fatigue behaviour of welded aluminium under multiaxial loading, Fatigue Fract. Engng Mater. Struct., Vol. 26, 2003, pp. 507–513
87. Kueppers M., Sonsino C.M.: Critical plane approach for the assessment of the fatigue behaviour of welded aluminium under multiaxial spectrum loading, in: Proceedings of 7th ICBMFF, Berlin 2004, pp. 373–380
88. Kueppers M., Sonsino C.M.: Festigkeitsverhalten von Aluminiumschweißverbindungen unter mehrachsigen Spannungszuständen, AiF-No. 10731N, LBF-Report No280542, Fraunhofer-Institut für Betriebsfestigkeit (LBF), Darmstadt 2001
89. Labesse-Jied F., Lebrun B., Petitpas E., Robert J.L.: Multiaxial fatigue assessment of welded structures by local approach, in: Biaxial/Multiaxial Fatigue Fracture, A. Carpinteri, M. de Freitas, A. Spagnoli, eds., Elsevier 2003, pp. 43–62
90. Lachowicz C., Łagoda T., Macha E.: Analityczne i algorytmiczne metody oceny trwałości zmęczeniowej przy obciążeniach losowych na przykładzie stali 10HNAP, XVI Sympozjum. Zmęczenie i Mechanika Pękania Materiałów i Konstrukcji, Akademia Techniczno-Rolnicza, Bydgoszcz 1996, ss. 113–122
91. Lachowicz C., Łagoda T., Macha E.: Comparison of analytical and algorithmical methods for life time estimation of 10HNAP steel under random loadings, Fatigue 96, Berlin 1996, G. Lutjering and H. Nowack, eds., Vol. I, pp. 595–600

92. Lachowicz C.T., Łagoda T., Macha E.: Covariance between components of biaxial stress state in fatigue life calculations, Mat. –wiss. u. Werkstofftech., Vol. 23, 1992, pp. 201–212
93. Lachowicz C.T., Łagoda T., Macha E.: Rola kowariancji składowych dwuosiowego stanu naprężenia przy obliczaniu trwałości zmęczeniowej, w: Wieloosiowe Zmęczenie Losowe Elementów Maszyn i Konstrukcji – część I, w: W. Będkowski i inni, Studia i Monografie z 63, WSI w Opolu, 1993, ss. 27–68
94. Lachowicz C., Łagoda T., Macha E.: Trwałość zmęczeniowa elementów maszyn ze stali 10HNAP w warunkach jednoosiowego obciążenia losowego, Problemy Maszyn Roboczych, Vol. 5, 1995, ss. 139–170
95. Lachowicz C., Łagoda T., Macha E., Dragon A., Petit J.: Selections of algorithms for fatigue life calculation of elements made of 10HNAP steel under uniaxial random loadings, Studia Geotechnika et Mechanica, Vol. XVIII, No 1–2, 1996, pp. 19–43
96. Lahti K.E., Hänninen H., Niemi E.: Nominal stress range fatigue of stainless steel fillet welds – the effect of weld size, J. Constructional Steel Research, Vol. 54, 2000, pp. 161–172
97. Lanning D.B., Nicholas T., Haritos G.K.: On the use of critical distance theories for prediction of the high cycle fatigue limit stress in notched Ti-6Al-4V, Int. J. Fatigue, Vol. 27, 2005, pp. 45–57
98. Lawrence F.W., Mattos R.J., Higashida Y., Burk J.D.: Estimating the fatigue crack initiation life of welds, ASTM STP 648, Fatigue Testing of Weldements, Philadelpia PA, ASTM, 1978, pp. 134–158
99. Liu J., Zenner H.: Berechnung der Dauerschwingfestigkeit bei mehrachsiger Beanspruchung, Mat.-wiss. u. Werkstofftech., Vol. 24, 1993, ss. 240–249
100. Lomolino S., Tovo R., dos Santos J.: On the fatigue behaviour and design curves of friction stir butt-welded Al alloys, Int. J. Fatigue, Vol. 27, 2005, pp. 305–316
101. Łagoda T.: Energetyczne modele oceny trwałości zmęczeniowej materiałów konstrukcyjnych w warunkach jednoosiowych i wieloosiowych obciążeń losowych, Studia i Monografie, Politechnika Opolska, Z.121, 2001, s. 148
102. Łagoda T.: Energetyczny model oceny trwałości zmęczeniowej elementów maszyn i konstrukcji poddanych jednoosiowemu losowemu rozciąganiu – część I – opracowanie modelu, Zakopane 2001, Zeszyty Naukowe Politechniki Opolskiej, Seria: Mechanika Z.64, Nr 265/2001, ss. 261–268
103. Łagoda T.: Energetyczny model oceny trwałości zmęczeniowej elementów maszyn i konstrukcji poddanych jednoosiowemu losowemu rozciąganiu – część II – weryfikacja modelu, Zakopane 2001, Zeszyty Naukowe Politechniki Opolskiej, Seria: Mechanika Z.64, Nr 265/2001, ss. 269–276
104. Łagoda T.: Energetyczne modele oceny trwałości zmęczeniowej materiałów konstrukcyjnych w warunkach jednoosiowych i wieloosiowych obciążeń losowych, Studia i Monografie, Politechnika Opolska, Z.121, 2001, s. 148

105. Łagoda T.: Energy models for fatigue life estimation under random loading – Part I – The model elaboration, Int. J. Fatigue, 2001, Vol. 23, 2001, pp. 467–480
106. Łagoda T.: Energy models for fatigue life estimation under random loading – Part II – Verification of the model, Int. J. Fatigue, 2001, Vol. 23, pp. 481–489
107. Łagoda T.: Metody prezentacji badań zmęczeniowych przy obciążeniach losowych, XXI Sympozjon PKM, Ustroń 2003, WNT, red. J. Wojnarowski i J. Drewniak, Tom II, 2003, ss. 81–90
108. Łagoda T.: Modelowanie lokalnych naprężeń i odkształceń w przekroju pręta gładkiego i z karbem przy kombinacji zginania ze skręcaniem, XIX Sympozjum Zmęczenia i Mechaniki Pękania, Bydgoszcz 2002, ss. 225–232
109. Łagoda T.: Rola związków korelacyjnych między składowymi stanu naprężenia przy obliczaniu trwałości zmęczeniowej tworzyw konstrukcyjnych, Praca doktorska, Raport 6/95, Wydział Mechaniczny WSI w Opolu, 1995, s. 184
110. Łagoda T.: Stress and strain distribution modelling in the bar section including stress gradients, Journal of Transdisciplinary Systems Science – Systems, Vol. 9, No. I, 2004, pp. 77–85
111. Łagoda T.: Uwagi o wybranych kryteriach naprężeniowych, VIII Krajowa Konferencja Mechaniki Pękania, Kielce/Cedzona 2001, Zeszyty Naukowe Politechniki Świętokrzyskiej, Mechanika 73, ss. 375–381
112. Łagoda T., Kaufmann H., Sonsino C.M.: Trwałość zmęczeniowa wybranych połączeń spawanych w warunkach stało- i zmiennoplitudowych obciążeń, Problemy Maszyn Roboczych, Z.20, 2002, ss. 119–126
113. Łagoda T., Küppers M.: Applying energy based criteria for calculation of an equivalent local stress amplitude of welded aluminium joints under in- and out-of-phase bending with torsion, Proceedings of the Seventh International Conference on Biaxial/Multiaxial Fatigue and Fracture, DVM, Berlin 2004, pp. 393–399
114. Łagoda T., Küppers M.: Trwałość zmęczeniowa aluminiowych złączy spawanych poddanych stałoamplitudowemu zgodnemu i niezgodnemu w fazie zginaniu ze skręcaniem, Czasopismo Techniczne – Mechanika z.1-M/2005, Wydawnictwo Politechniki Krakowskiej, Kraków 2005, ss. 265–274
115. Łagoda T., Macha E.: A Review of High-Cycle Fatigue Models under Non-Proportional Loadings, in: Fracture from Defects, Proceedings of ECF-12, Sheffield, M.W. Brown, E.R. de los Rios and K.J. Miller, eds., EMAS 1998, Vol. I, pp. 73–78.
116. Łagoda T., Macha E.: Assessment of long-life time under uniaxial and biaxial random loading with energy parameter on the critical plane, Fatigue'99, Proceedings of the seventh Intern. Fatigue Congress, X.-R. Wu and Z.-G. Wang, eds., Vol. II, pp. 965–970

117. Łagoda T., Macha E.: Energetic approach to fatigue under combined bending with torsion of smooth and notched specimens, presented in 3rd annual meeting of Copernicus Contract CIPA CT940194 and 6th International Scientific Conference „Achievements in the Mechanical and Materials Engineering, 1–3 December 1997, Miscolc – Hungary, Proceedings, T. Laszlo, ed., Miscolc 1998, pp. 53–64
118. Łagoda T., Macha E.: Energetyczny model oceny trwałości zmęczeniowej elementów maszyn poddanych kombinacji zginania ze skręcaniem, XIII Konferencja Naukowa – Problemy Rozwoju Maszyn Roboczych – Zakopane 2000, Politechnika Łódzka Filia w Bielsku-Białej, ss. 285–292
119. Łagoda T., Macha E.: Energetyczny opis zmęczenia cyklicznego próbek gładkich i z karbem z uwzględnieniem gradientów naprężeń przy kombinacji zginania ze skręcaniem, XVII Sympozjon Zmęczenia Materiałów i Konstrukcji, Bydgosz-Pieczyska 1998, ss. 183–188
120. Łagoda T., Macha E.: Energy approach to fatigue life estimation under combined tension with torsion, 7 Letnia Szkoła Mechaniki Pękania – Pokrzywna, Zeszyty Naukowe Politechniki Opolskiej, Z.67, Nr269/2001, Opole 2001, pp. 163–182
121. Łagoda T., Macha E.: Energy approach to fatigue under combined cyclic bending with torsion of smooth and notched specimens, Physicochemical Mechanics of Materials, No 5, 1998, pp. 34–42
122. Łagoda T., Macha E.: Energy – based approach to damage cumulation in random fatigue, in: Reliability Assessment of Cyclically Loaded Engineering Structures, R.A. Smith, ed., Kluwer Academic Publishers, 1997, pp. 435–442
123. Łagoda T., Macha E.: Fatigue lives under biaxial random loading according to normal stress, strain and strain energy density in the critical plane, Life Assessment and Management for Structural Components, Proceedings of the Conference, ed., V.T. Troshenko, Kijów 2000, Tom 1, pp. 119–124
124. Łagoda T., Macha E.: Generalization of energy multiaxial cyclic fatigue criteria to random loadings, Multiaxial Fatigue and Deformation: Testing and Prediction, ASTM STP 1387, S. Kalluri and P.J. Bonacuse, eds., American Society for Testing and Materials, West Conshohocken, PA, 2000, pp. 173–190
125. Łagoda T., Macha E.: Gęstość energii odkształceń w płaszczyźnie krytycznej jako parametr wieloosiowego zmęczenia, III Konferencja PTMTS – Nowe Kierunki Rozwoju Mechaniki, Zeszyty Naukowe Katedry Mechaniki Stosownej, nr 14, Politechnika Śląska, Gliwice 2000, ss. 95–100
126. Łagoda T., Macha E.: Influence of cross-correlation between normal stresses on biaxial fatigue life, 11th European Conference on Fracture, Futuroscope 1996, J. Petit, ed., Vol. II, pp. 1361–1366
127. Łagoda T., Macha E.: Modele trwałości zmęczeniowej w zakresie dużej liczby cykli przy obciążeniach nieproporcjonalnych, XVIII Sympozjon PKM, Kielce-Ameliówka 1997, Vol. II, ss. 225–230

128. Łagoda T., Macha E.: Simulation of cross correlation effect on biaxial random fatigue, in: Fatigue 93, J.P. Bailon and J.I. Dickson, eds., EMAS (U.K.), 1993, Vol. III, pp. 1539–1544
129. Łagoda T., Macha E.: Trwałość zmęczeniowa przy dwuosiowych nieproporcjonalnych obciążeniach losowych według naprężenia, odkształcenia i energii właściwej odkształceń normalnych w płaszczyźnie krytycznej, Problemy Maszyn Roboczych, z. 15, 2000, ss. 47–69
130. Łagoda T., Macha E.: Trwałość zmęczeniowa przy kombinacji rozciągania ze skręcaniem w ujęciu energetycznym, XX Sympozjon Podstaw Konstrukcji Maszyn, Zeszyty Naukowe Politechniki Opolskiej, Z.69, Nr 271/2001, Opole 2001, ss. 171–182
131. Łagoda T., Macha E.: Uogólnienie energetycznych modeli oceny wieloosiowego zmęczenia na zakres obciążeń losowych, VII Krajowa Konferencja Mechaniki Pękania, Kielce-Cedzyna, Zeszyty Naukowe Politechniki Świętokrzyskiej – Mechanika 62, 1999, tom. II, ss. 17–24
132. Łagoda T., Macha E.: Wieloosiowe zmęczenie losowe elementów maszyn i konstrukcji – część II, Studia i Monografie, Z 76, WSI Opole, Opole 1995, s. 136
133. Łagoda T., Macha E.: Wieloosiowe zmęczenie losowe elementów maszyn i konstrukcji – cz. III, Studia i Monografie, Z.104, Politechnika Opolska, Opole 1998, s. 196
134. Łagoda T., Macha E.: Wyznaczanie trwałości zmęczeniowej w warunkach nieproporcjonalnych obciążeń losowych za pomocą parametru naprężeniowego, odkształceniowego i energetycznego w płaszczyźnie krytycznej, XVIII Sympozjon Zmęczenia Materiałów i Konstrukcji, Akademia Techniczno-Rolnicza, Bydgoszcz 2000, ss. 281–288
135. Łagoda T., Macha E., Achtelik H., Karolczuk A., Niesłony A., Pawliczek R.: Wieloosiowe zmęczenie losowe elementów maszyn i konstrukcji-część IV-Trwałość zmęczeniowa z uwzględnieniem gradientów naprężeń w ujęciu energetycznym, Studia i Monografie, Z.139, Politechnika Opolska, Opole 2002, s. 112
136. Łagoda T., Macha E., Będkowski W.: A critical plane approach based on energy concepts: Application to biaxial random tension-compression high-cycle fatigue regime, Int. J. Fatigue, Vol. 21, 1999, pp. 431–443
137. Łagoda T., Macha E., Będkowski W.: Wyznaczanie trwałości zmęczeniowej stali 10HNAP w jedno- i dwuosiowym losowym stanie naprężenia za pomocą parametru energetycznego, VII Krajowa Konferencja Mechaniki Pękania, Kielce-Cedzyna, Zeszyty Naukowe Politechniki Świętokrzyskiej – Mechanika 62, 1999, tom. II, ss. 9–16
138. Łagoda T., Macha E., Dragon A., Petit J.: Influence of correlations between stresses on calculated fatigue life on machine elements, Int. J. Fatigue, Vol. 18, 1996, pp. 547–555

139. Łagoda T., Macha E., Molski K., Ferenc R.: Rozkłady naprężeń i odkształceń dla próbek gładkich i z karbem w przypadku kombinacji zginania ze skręcaniem, VIII Krajowa Konferencja Mechaniki Pękania, Kielce/Cedzyna 2001, Zeszyty Naukowe Politechniki Świętokrzyskiej, Mechanika 73, ss. 55–62
140. Łagoda T., Macha E., Niesłony A., Morel F.: Estimation of the fatigue life of high strength steel under variable-amplitude tension with torsion with use of the energy parameter in the critical plane, In: Biaxial/Multiaxial Fatigue and Fracture, A. Carpinteri, M. DeFreitas and A. Spagnoli, eds., Elsevier 2003, pp. 183–202
141. Łagoda T., Macha E., Niesłony A., Morel F.: Estymowanie trwałości zmęczeniowej stali 35NCD16 przy kombinacji rozciągania ze skręcaniem za pomocą parametru energetycznego, XVIII Sympozjon Zmęczenia Materiałów i Konstrukcji, Bydgoszcz-Pieczyska, maj 2000, ss. 289–296
142. Łagoda T., Macha E., Niesłony A., Morel F.: The energy approach to fatigue life of high strength steel under variable-amplitude tension with torsion, proceedings of 6th ICBMFF, Lisbon 2001, M. M. de Freitas, ed., Vol. I, pp. 233–240
143. Łagoda T., Macha E., Niesłony A., Muller A.: Comparison of calculation and experimental fatigue lives of some chosen cast irons under combined tension and torsion, The 13th European Conference on Fracture Mechanics: Application and Challenges, ECF13, 6–9th September 2000, San Sebastian, Spain, eds., M. Fuentec, M. Elices, A. Martin-Meizoso and J.M. Martinez-Esnado, Elsevier, Abstract Volume 256 pp, CD-Rom, ps 6
144. Łagoda T., Macha E., Niesłony A., Muller A.: Fatigue life of cast irons GGG40, GGG60 and GTS45 under combined variable amplitude tension with torsion, The Archive of Mechanical Engineering, Vol. XLVIII, No. 1, 2001, pp. 55–69
145. Łagoda T., Macha E., Niesłony A., Muller A.: Trwałość zmęczeniowa żeliw GGG40, GGG60 I GTS45 przy kombinacji losowego rozciągania ze skręcaniem, Przegląd Mechaniczny, Vol. 5–6, 2000, ss. 28–32
146. Łagoda T., Macha E., Pawliczek R.: Fatigue life of 10HNAP steel under random loading with mean value, Int. J. Fatigue, Vol. 23, 2001, pp. 283–291
147. Łagoda T., Macha E., Pawliczek R.: Obliczanie trwałości zmęczeniowej przy jednoosiowym losowym obciążeniu z udziałem wartości średnich, XVII Sympozjon Zmęczenia Materiałów i Konstrukcji, Bydgosz-Pieczyska 1998, ss. 189–194
148. Łagoda T., Macha E., Pawliczek R.: Wpływ wartości oczekiwanej losowej historii naprężenia na trwałość zmęczeniową stali 10HNAP przy jednoosiowym rozciąganiu, Problemy Maszyn Roboczych, z.11, 1998, ss. 121–138
149. Łagoda T., Macha E., Sakane M.: Correlation of biaxial low-cycle fatigue lives of SUS304 stainless steel with energy parameter in critical plane at 923 K, 6th ISCCP – Białowieża, 1998, A. Jakowluk and Z. Mróz, eds., pp. 343–356

150. Łagoda T., Macha E., Sakane M.: Estymowanie trwałości zmęczeniowej elementów maszyn ze stali nierdzewnej w podwyższonej temperaturze za pomocą parametru energetycznego w płaszczyźnie krytycznej, Problemy Maszyn Roboczych, z.12, 1998, ss. 61–77
151. Łagoda T., Macha E., Sakane M.: Estimation of fatigue lifetime of SUS304 steel under 923K with the energy parameter in the critical plane, J. Theor. Appl. Mech., Vol. 41, 2003, pp. 55–73
152. Łagoda T., Macha E., Sakane M.: Opis zmęczenia elementów maszyn za pomocą energii właściwej odkształceń w płaszczyźnie krytycznej, Sympozjon PKM, Zielona Góra-Świnoujście, Politechnika Zielonogórska 1999, T. II, ss. 67–72
153. Łagoda T., Marciniak Z.: Wyznaczanie lokalnych naprężeń i odkształceń sprężystoplastycznych przy zginaniu ze skręcaniem pręta gładkiego, XIX Sympozjum Zmęczenia i Mechaniki Pękania, Bydgoszcz 2002, ss. 233–240
154. Łagoda T., Niesłony A., Macha E., Morel F.: Trwałość zmęczeniowa stali 35NCD16 przy kombinacji losowego rozciągania ze skręcaniem w ujęciu energetycznym, Przegląd Mechaniczny, Nr 5, 2004, ss. 25–31
155. Łagoda T., Ogonowski P.: Fatigue life of AlCu4Mg1 aluminium alloy under constant-amplitude in- and out-of-phase bending with torsion, ICF Turyn 2005, 20–25 March 2005, ed., A. Carpinteri, Politecnico di Torino, CD, ps 6
156. Łagoda T., Ogonowski P.: Kryteria wieloosiowego zmęczenia losowego oparte na naprężeniowych, odkształceniowych i energetycznych parametrach uszkodzenia w płaszczyźnie krytycznej, Przegląd Mechaniczny, Nr 7–8, 2004, ss. 32–40
157. Łagoda T., Sonsino C.M.: Comparison of different methods for presenting constant and variable amplitude loading fatigue results, Mat.-wiss.u. Werstofftech., Vol. 35, 2004, pp. 13–20
158. Łagoda T., Sonsino C.M.: Trwałość zmęczeniowa wybranych złączy spawanych według wybranych kryteriów energetycznych, Problemy Maszyn Roboczych, Z.22, 2003, ss. 47–58
159. Łagoda T., Sonsino C.M.: Wyznaczanie współczynnika działania karbu na podstawie fikcyjnego promienia karbu, Przegląd Mechaniczny, Nr 1, 2004, ss. 23–25
160. Łagoda T., Sonsino C.M., Kaufmann H.: Cumulative fatigue damage of welded high-strength steels under uniaxial constant and variable amplitude loading, Working Group meeting of the International Institute of Welding (IIW) within the International Conference, Copenhagen, June 2002 (oral presentation)
161. Macha E.: Modele matematyczne trwałości zmęczeniowej materiałów w warunkach losowego złożonego stanu naprężenia, Prace Nauk. Inst. Mater. i Mech. Tech. Pol. Wrocł. Nr 41, Seria: Monografie Nr 13, Wrocław 1979, s. 99
162. Macha E.: Simulation investigations of the position of fatigue fracture plane in materials with biaxial loads, Mat. –wiss. U. Werkstofftech. Nr 20, 1989, Teil I, Heft 4/89, pp. 132–136, Teil II, Heft 5/89, pp. 153–163

163. Macha E.: Simulation of fatigue process in material subjected to random complex state of stress, in: Simulation of Systems, L. Dekker ed., North – Holland Publishing Company, Amsterdam 1976, pp. 1033–1041
164. Macha E., Sonsino C.M.: Energy criteria of multiaxial fatigue failure, Fatigue Fract. Engng Mater. Struct., Vol. 22, 2000, pp. 1053–1070
165. Maddox S.J., Sonsino C.M.: Multiaxial Fatigue of Welded Structures: Problems and Present Solutions, In: Proc. Of the Sisth International conference on Biaxial/Multiaxial Fatigue and Fracture, ed., Manuel de Freitas, Vol. II, Lisboa 2001, pp. 3–16
166. Maddox S.J.: Review of fatigue assessment procedures for welded aluminium structures, Int. J. Fatigue, Vol. 25, 2003, pp. 1359–1378
167. Markusik S., Łukasik T.: Naprężenia zmęczeniowe w złączach spawanych konstrukcji stalowych dźwignic obciążonych dynamicznie, Transport Przemysłowy, No. 4(6), 2001, ss. 5–8
168. Mayer H., Ede C., Allison J.E.: Influence of cyclic loads below endurance limit or threshold stress intensity on fatigue damage in cast aluminium alloy 319-T7, Int. J. Fatigue, Vol. 27, 2005, pp. 129–141
169. Miner M.A.: Cumulative damage in fatigue, J. Applied Mechanics, Vol. 12, 1945, pp. 159–164
170. Mueller R.A., Georg D.D., Johnson G.R.: A random number generator for microprocessors, Simulation, 1977, pp. 123–127
171. Nadot Y., Denier V.: Fatigue failure at suspension arm: experimental analysis and multiaxial criterion, Engineering Failure Analysis, Vol. 11, 2004, pp. 485–499
172. N'diaye A., Azari Z., Pluvinage G., Hariri S.: Stress concentration factor analysis for notche welded tubular T-joints, Int. J. Fatigue, Vol. 29, No. 8, 2007, pp. 1554–1570
173. Neuber H.: Kerbspannungslehre – Theorie der Spannungskonzentration, Genaue Berechnung der Festigkeit Springer. Verlag, Berlin 1985. 3rd edition
174. Neuber H.: Über die Berücksichtigung der Spannungskonzentration bei Festigkeitsberechnungen, Konstruktion, Heft 7, 1968, pp. 245–251
175. Nie H., Wu F.M., Liu J.F.: A variable K_f – Neuber's rule for predicting fatigue crack initiation life, Fatigue Fract. Engng Mater. Struct., Vol. 17, 1994, pp. 1015–1023
176. Niemi E.: Aspects of good design practice for fatigue-loaded welded components, Fatigue Design, ESIS 16, J. Solin, G. Marquis, A. Siljander and S. Sipila, eds., Mechanical Engineering Publications, London 1993, pp. 333–351
177. Niemi E.: Structural stress approach to fatigue analysis of welded components – designer's guide, IIIW-Doc. XIII-1819-00/XV-1091-01 (Final Draft), International Institute of Welding, 2001
178. Niemi E.: Stress determination for fatigue analysis of welded components, Abington, Cambridge, International Institute of Welding, Ambington Publishing, 1995

179. Niesłony A., Kardas D., Łagoda T., Macha E.: Parametr energetyczny w ocenie trwałości zmęczeniowej przy zmiennoamplitudowym rozciąganiu-ściskaniu stali 12010.3, Zmęczenie i Mechanika Pękania, Akademia Techniczno-Rolnicza, Bydgoszcz 2004, ss. 273–279
180. Niesłony A., Łagoda T., Macha E.: Comparison of the rain-flow algorithm and the spectral method for fatigue life determination under multiaxial random loading, Symposium on fatigue testing and analysis under variable amplitude loading conditions, Tours 2002, Extended abstracts, pp. 40(1)–(2)
181. Nisitani H.: Stress concentration of a strip with double edge notches under tension or in-plane bending, Engng Fract. Mech., Vol. 23, 1986, pp. 1051–1065
182. Noda N.A., Sera M., Takasa Y.: Stress concentration factors for round and flat test specimens with notches, Int. J. Fatigue, Vol. 17, 1995, pp. 163–178
183. Noda N.A., Takasa Y.: Stress concentration formula useful for any dimensions of shoulder fillet in a round bar under tension and bending, Fatigue Fract. Engng Mater. Struct., Vol. 26, 2003, pp. 245–255
184. Noda N.A., Takasa Y.: Stress concentration formula useful for any shape of notch in a round test specimen under tension and under bending, Fatigue Fract. Engng Mater. Struct., Vol. 22, 1999, pp. 1071–1082
185. Noda N.A., Tsubaki M.A., Nisitani H.: Stress concentration of a strip with V- or U-shaped notches under transverse bending, Engineering Fracture Mechanics, Vol. 31, 1988, pp. 119–133
186. Olivier R., Amstutz H.: Fatigue strength of shear loaded welded joints according to the local concept, Materialwissenschaft und Werkstofftechik, Vol. 4, 2001, pp. 287–297
187. Ogonowski P., Kardas D., Kluger K., Łagoda T.: Weryfikacja energetycznego parametru uszkodzenia w przypadku zginania ze skręcaniem, IX Krajowa Konferencja Mechaniki Pękania. Kielce 2003, Politechnika Świętokrzyska, ss. 355–364
188. Ogonowski P., Łagoda T.: Energetyczny parametr uszkodzenia w złożonym stanie obciążenia oparty na płaszczyźnie krytycznej, IX Krajowa Konferencja Mechaniki Pękania. Kielce 2003, Politechnika Świętokrzyska, ss. 365–372
189. Ogonowski P., Łagoda T.: Parametr uszkodzenia w ujęciu energetycznym w złożonym stanie naprężenia w przypadku występowania spiętrzenia naprężeń, Zmęczenie i Mechanika Pękania, Akademia Techniczno-Rolnicza, Bydgoszcz, 2004, ss. 289–296
190. Ogonowski P., Łagoda T.: Trwałość zmęczeniowa spieków z koncentratorami naprężeń według parametru energetycznego, XXII Sympozjon PKM, Jurata 2005
191. Ogonowski P., Łagoda T.: Wyznaczania trwałości zmęczeniowej z uwzględnieniem odkształceń plastycznych, III Sympozjum Mechaniki Zniszczenia Materiałów i Konstrukcji, Augustów, 1 – 4 czerwca 2005, Politechnika Białostocka, ss. 275–278

192. Ogonowski P., Łagoda T., Achtelik H.: Trwałość zmęczeniowa stali 10HNAP z koncentratorami naprężeń w warunkach proporcjonalnego zginania ze skręcaniem, Problemy Maszyn Roboczych, Z.24, 2004, ss. 75–84
193. Ogonowski P., Łagoda T., Karolczuk A.: Weryfikacja kryterium wyznaczania trwałości zmęczeniowej z uwzględnieniem odkształceń plastycznych, III Sympozjum Mechaniki Zniszczenia Materiałów i Konstrukcji, Augustów, 1 – 4 czerwca 2005, Politechnika Białostocka, ss. 279–284
194. Osiński Z., Bajon W., Szucki T.: Podstawy Konstrukcji Maszyn, PWN, Warszawa 1975, s. 471
195. Palin-Luc T.: Fatigue multiaxiale d'une fonte GS sous combinescions combines d'amplitude variable, These de Docteur, ENSAM Bordeaux 1996
196. Palmgren A.: Die Lebensdauer Von Kugellagern, VDI-Z, Vol. 68, 1924, ss. 339–341
197. Pluvinage G.: Notched effect in high cycle fatigue, ICF9, pp. 1239–1250
198. Polge R.J., Holliday E.M., Bhagavan B.K.: Generation of a pseudo-random set with desired correlation and probability distribution, Simulation, 1973, pp. 153–158
199. Porębska M., Skorupa A.: Połączenia spójnościowe, PWN, Warszawa 1997, s. 217
200. Qilafku G., Pluvinage G.: Multiaxial fatigue criterion for notched specimens including the effective stress range, relative stress gradient, and hydrostatic pressure, Materials Science, Vol. 37, 2001, pp. 573–582
201. Qylafku G., Azari Z., Gjonaj M., Pluvinage G.: On the fatigue failure and life prediction of the notched specimens, Physicochem. Mech. Mater., No. 5, 1998, pp. 17–26
202. Radaj D., Sonsino C.M.: Fatigue Assessment of Welded Joints by Local Approaches, Abington Publishing, Cambridge, 1998
203. Rosochowicz K.: Badania eksperymentalne procesów zmęczenia konstrukcji kadłubów statków, W: Metody doświadczalne w zmęczeniu materiałów i konstrukcji. Badania konstrukcji, red. J. Szali, Wyd. Uczel ATR w Bydgoszczy, Bydgoszcz, 2000, ss. 5–100
204. Rykaluk K.: Pęknięcia w konstrukcjach stalowych, Dolnośląskie Wydawnictwo Edukacyjne, Wrocław 2000, s. 226
205. Schütz W., Klätschke H., Hück M., Sonsino C.M.: Standardized load sequence for offshore structures – Wash 1, Fatigue Fract. Engng Mater. Struct., Vol. 13, 1990, pp. 15–29
206. Schütz W., Klätschke H., Steinhilber H., Heuler P.: Standardized load sequence for wheel suspension components – CARLOS, LBF Report No. FB-191, Darmsadt 1990
207. Sedlacek G.: Eurocode 3: Unified European rules for the design of steel structures, Welding in the World, Lo.39, 1997, pp. 8–15
208. Seeger T.: Stahlbauhandbuch – Band 1, Teil B, Abschnitt "Grundlagen für Betriebsfestigkeitsnachweise". Stahlbau-Verlagsgesellschaft mbH, Düsseldorf, 1996
209. Serensen S.V., Kogayev V.P., Snajdorovic R.M.: Nesuščaja sposobnost i rasčet detalej mašin na pročnost. Izd. 3-e, Mašinostroenie 1975, s. 488

210. Sharpe W.N.: ASME 1993 Nadai lecture – elstoplastic stress and strain conecentration, J. Engng Mater. Technol., Vol. 117, 1995, pp. 1–7
211. Shinozuka M.: Simulation of multivariate and multidimensional random processes, J. Acoust. Soc. Am., Vol. 49, part 2, 1977, pp. 357–367
212. Słowik J., Łagoda T.: Wyznaczanie odkształceń w dnie karbu obrączkowego dla wybranych materiałów, XXII Sympozjon PKM, Jurata 2005
213. Słowik J., Łagoda T.: Analiza sprężysto-plastycznych odkształceń w elementach z karbem obrączkowym przy jednoosiowym rozciąganiu-ściskaniu, III Sympozjum Mechaniki Zniszczenia Materiałów i Konstrukcji, Augustów, 1 – 4 czerwca 2005, Politechnika Białostocka, ss. 385–388
214. Smith K.N., Watson P., Topper T.H.: A stress-strain function for the fatigue of metals, J. Mater., Vol. 5, 1970, pp. 767–776
215. Sobczykiewicz W.: Wymiarowanie spawanych konstrukcji nośnych maszyn z zakresie trwałości zmęczeniowej, Prace naukowe, Mechanika, Z.157, Politechnika Warszawska, Warszawa, 1994, s. 88
216. Sobczykiewicz W.: Wymiarowanie spawanych konstrukcji nośnych w zakresie zjawiska zmęczenia. Zasady ogólne, W: Metody doświadczalne w zmęczeniu materiałów i konstrukcji. Badania konstrukcji, red. J. Szali, Wyd. Uczel ATR w Bydgoszczy, Bydgoszcz 2000, ss. 157–210
217. Sonsino C.M.: Fatigue Behaviour of Welded Components under Complex Elasto-Plastic Multiaxial Deformations, Eur-Report No. 16024, Luxemburg, 1997
218. Sonsino C.M.: Private correspondence, 2001–03
219. Sonsino C.M.: Methods to determine relevant material properties for the fatigue design of powder metallurgy parts, Powder Metallurgy International, Vol. 16, Nr. 1 ss. 34–38, Nr. 2, ss. 73–77
220. Sonsino C.M.: Multiaxial Fatigue of Welded Joints Under In-Phase and Out-of-Phase Local Strain and Stresses, Int. J. Fatigue, Vol. 17, 1995, pp. 55–70
221. Sonsino C.M.: Overview of the State of the Art on Multiaxial Fatigue of Welds. In: Multiaxial Fatigue and Fracture, E. Macha, W. Będkowski and T. Łagoda, eds., ESIS Publ., No. 25, 2000, pp. 195–217
222. Sonsino C.M.: Zur bewrtung des schwingfestigkeitsverhaltens von bauteilen mit hilfe örtlicher beanspruchungen, Sonderkdruck aus Konstruktion, Vol. 45, H. 1, 1993, ss. 25–33
223. Sonsino C.M., Grubisic V.: Requirements for operational fatigue strength of high quality cast components, Mat.-wiss., u. Werkstofftech., Vol. 27, 1996, pp. 373–390
224. Sonsino C.M., Kaufmann H., Demofonti G., Riscifuli. S., Sedlacek G., Müller C., Hanus F., Wegmann H.G.: High-strength steels in welded state for light-weight constructions under high and variable stress peaks, ECSC steel research programme, CSM – Roma, LBF – Darmstadt, 1999, will be published by the European Commission, Brussels
225. Sonsino C.M., Kaufmann H., Hanselka H.: Cumulative damage of laser-beam welded thin automotive hollow structures of steel and aluminium, ICEM12, Politechnico di Bari, 2004, CD, p. 7

226. Sonsino C.M., Kueppers M.: Fatigue behaviour of welded aluminium under multiaxial loading, in: Proceedings of 6th ICBMFF, European Structural Integrity Society, ESIS, Lisbon 2001, pp. 57–64
227. Sonsino C.M., Łagoda T.: Assessment of multiaxial fatigue behaviour of welded joints under bending and torsion by application of a fictitious radius, Int. J. Fatigue, Vol. 26, 2004, pp. 265–279
228. Sonsino C.M., Łagoda T.: Damage accumulation under variable amplitude loading of welded medium- and high-strength steels, Int. J. Fatigue, Vol. 26, No. 5, 2004, pp. 487–495
229. Sonsino C.M., Morgenstern C., Wohlfahrt H., Krull P., Nitschke-Pagel T., Dilthey K., Kessel M., Krüger P., Gebur J., Kueppers M.: Grundlagen für den Leichtbau energiesparender Nutzfahrzeuge auf Basis neuartiger Schweiß- und Auslegungsverfahren für Aluminiumkonstruktionen – Schweißverfahren und Schwingfestigkeit BMBF (MATECH)-Projekt, Förderkennzeichen 03N30479, Fraunhofer-Institut für Betriebsfestigkeit (LBF), Darmstadt LBF-Bericht Nr. 8723, 2002
230. Sonsino C.M., Radaj D., Brandt U., Lehrke H.P.: Fatigue Assessment of Welded Joints in AlMg4.5Mn Aluminium Alloy (AA 5083) by Local Approaches, Int. J. Fatigue, Vol. 21, 1999, pp. 985–999
231. Stromeyer C.E.: The determination of fatigue limits under alternating stress conditions, Proc. R. Soc. London, Ser. A, Vol. 90, 1914, pp. 411–425
232. Susmel L., Tovo R.: Modified Wöhler curve method and Eurocode 3: Accuracy in predicting the multiaxial fatigue strength of welded joints, pp. 203–207
233. Susmel L., Tovo R.: On the use of nominal stresses to predict the fatigue strength of welded joints under biaxial cyclic loading, Fatigue Fract. Engng Mater. Struct., Vol. 27, 2004, pp. 1005–1024
234. Śledziewski E.: Projektowanie stalowych konstrukcji spawanych, WNT, Warszawa 1972, s. 409
235. Szala J.: Hipotezy sumowania uszkodzeń zmęczeniowych, ATR Bydgoszcz 1998, s. 175
236. Thesen A.: An efficient generator of uniformy distributed random varieties between zero and one, Simulation, Vol. 44, 1985, pp. 17–22
237. Trwałość zmęczeniowa elementów maszyn i konstrukcji w warunkach wieloosiowych obciążeń losowych, Prace Nauk. CPBP 02.05, Wyd. Pol. Warszaw., Warszawa, 1990, (pod red. E. Machy) s. 93
238. Watson P., Dabell B.J.: Cycle Counting and Fatigue Damage, J. Soc. Environ. Engi., September 1976, pp. 3–8
239. Werner S., Sonsino C.M., Radaj D.: Schwingfestigkeit von Schweißverbindungen aus der Aluminiumlegierung AlMg4,5Mn (AA5083) nach dem Konzept der Mikrostützwirkung, Materialwissenschaft und Werkstofftechnik 30, No. 3, 1999, pp. 125–135
240. Witt M.: Schwingfestigkeit von Schweissverbindungen bei zusammengesetzter Betriebsbeanspruchung, Dissertation, Technischen Univeristat Clausthal, Clausthal 2000, ss. 121–138

241. Witt M., Yousefi F., Mauch H., Zenner H.: Schwingfestigkeit von Schweißverbindugen bei mehrachsiger Beanspruchung. Versuche und Berechnungen, DVS Kolloquium Schweißkonstruktion, Braunschweig, 1997
242. Witt M., Yousefi F., Zenner H.: Fatigue strength of welded joints under multiaxial loading: comparison between experiments and calculations, In: S. Kalluri, ed., ASTM: Multiaxial fatigue and deformation: testing and prediction. West Conshohocken, Pennsylvania, American Society for Testing and Materials, 2000, ss. 191–210 (ASTM STP 1387)
243. Witt M., Zenner H., Yousefi F.: Fatigue strength of welded components under multiaxial random loading. Comparison of different lifetime prediction concepts, Proceedings of 6th International Conference on Biaxial/Multiaxial Fatigue and Fracture, Lisbon 2004, Vol. I, pp. 29–39
244. Wöhler A.: Z. Bauwwesen, Vol. 8, 1858, ss. 642–652
245. Wróbel J.: Symulacyjne badanie jakości w nieliniowej stochastycznej dynamice maszyn, Prace Naukowe, Mechanika, z. 92, Politechnika Warszawska, Warszawa 1985, s. 131
246. Xiao Z.G., Yamada K.: A method of determining geometric stress for fatigue strength evaluation of steel welded joint, Int. J. Fatigue, Vol. 26, 2004, pp. 1277–1293
247. Zhi-Zhong H., Shu-Zhen C.: Relationship between fatigue notch factor and strength, Engng Fract. Mech., Vol. 48, pp. 127–134
248. Zieliński A.: Generatory liczb losowych. Programowanie I testowanie maszyn cyfrowych, WNT, Warszawa, 1979, s. 157
249. Życzkowski M.: Strength of Structural Elements, Studies Applied Mechanics 26, PWN, Warszawa 1991, p. 774

Summary

The paper presents fatigue life calculations for some chosen welded joints. The results were verified on the basis of fatigue tests of steel and aluminium welded joints under uniaxial and multiaxial loading states. Uniaxial loading concern pure tension-compression and alternating bending under cyclic and random tests of specimens made of steel. One fatigue characteristic can be determined for four considered materials. From the calculations it appears that fatigue strength does not strongly depend on a type of the considered welded joint – it is more dependent on loading type (bending and axial loading). The calculated sum of damages for normal distribution (the Gaussian spectrum) and normal distribution with overloads is included into the scatter band with coefficient 3 according to the Palmgren-Miner hypothesis at the significance level 5%. In the case of the considered welded joints, the mean stress value does not influence the fatigue life. The same fatigue curves have been obtained for symmetric and pulsating loading.

Complex stress states concern loading under combined proportional and non-proportional cyclic bending with torsion. Moreover, for aluminium joints verification was also done under random loading. Evaluating the multiaxial fatigue histories in welded joints by local stresses and strains, we must know the actual local radius at the weld edge. Owing to the fictitious local radius, when – in the worst case – for sharp notches $\rho = 0$, it is possible to calculate coefficients of the notch action for bending, K_{fb} and for torsion, K_{ft}. In this order we must determine fictitious radii of the notch ρ_f for bending and for torsion. In the case of steel welded joints, these radii are $\rho_{fb} = 1.16$ mm for bending and $\rho_{ft} = 0.4$ mm for torsion. The normal and shear strain energy density parameter in the critical plane determined by the energy density parameter of normal and shear strain for steel welded joints gives comparable results. However, if the normal strain energy density parameter is assumed as the critical plane, it is necessary to determine, in an experimental way, the weight function including the shear strain energy density parameter in this plane. Thus, application of the energy criterion defined in the plane determined by the shear strain energy density parameter, is recommended. In the case of aluminium welded joints, satisfactory results of fatigue life calculations were obtained for the criterion of energy parameter in the plane defined by the shear strain energy density parameter. In the case of application of the energy criterion in the plane defined by the normal strain energy density parameter, the obtained calculated fatigue lives were strongly overestimated in comparison of the experimental results. Application of the maximum shear and normal strain energy density parameter in the critical plane for aluminium welded joints subjected to variable-amplitude bending with torsion seems to be right under the Palmgren-Miner hypothesis and application of the correction coefficient, like in the case of the Serensen-Kogayev hypothesis, which is the quotient of the weighed amplitude of the fatigue curve inclination in energy approach and the maximum amplitude in the history.